Practical Genetics for Aquaculture

Practical Genetics for Aquaculture

C. Greg Lutz

 Fishing News Books
An imprint of Blackwell Science

 Blackwell Science

Copyright © 2001

Fishing News Books
A division of Blackwell Science Ltd
Editorial Offices:
Osney Mead, Oxford OX2 0EL
25 John Street, London WC1N 2BS
23 Ainslie Place, Edinburgh EH3 6AJ
350 Main Street, Malden
 MA 02148 5018, USA
54 University Street, Carlton
 Victoria 3053, Australia
10, rue Casimir Delavigne
 75006 Paris, France

Other Editorial Offices:

Blackwell Wissenschafts-Verlag GmbH
Kurfürstendamm 57
10707 Berlin, Germany

Blackwell Science KK
MG Kodenmacho Building
7–10 Kodenmacho Nihombashi
Chuo-ku, Tokyo 104, Japan

Iowa State University Press
A Blackwell Science Company
2121 S. State Avenue
Ames, Iowa 50014–8300, USA

The right of the Author to be identified as
the Author of this Work has been asserted in
accordance with the Copyright, Designs and
Patents Act 1988.

All rights reserved. No part of this publication
may be reproduced, stored in a retrieval system,
or transmitted, in any form or by any means,
electronic, mechanical, photocopying, recording
or otherwise, except as permitted by the
UK Copyright, Designs and Patents Act 1988,
without the prior permission of the publisher.

First published 2001

Set in 10.5/12 pt Palatino
by Sparks Computer Solutions Ltd, Oxford
http://www.sparks.co.uk

The Blackwell Science logo is a
trade mark of Blackwell Science Ltd,
registered at the United Kingdom
Trade Marks Registry

DISTRIBUTORS
Marston Book Services Ltd
PO Box 269
Abingdon
Oxon OX14 4YN
(*Orders:* Tel: 01865 206206
 Fax: 01865 721205
 Telex: 83355 MEDBOK G)

USA and Canada
Iowa State University Press
A Blackwell Science Company
2121 S. State Avenue
Ames, Iowa 50014-8300
(*Orders:* Tel: 800-862-6657
 Fax: 515-292-3348
 Web: www.isupress.com
 email: orders@isupress.com)

Australia
Blackwell Science Pty Ltd
54 University Street
Carlton, Victoria 3053
(*Orders:* Tel: 03 9347 0300
 Fax: 03 9347 5001)

A catalogue record for this title
is available from the British Library

ISBN 0-85238-285-5

Library of Congress
Cataloging-in-Publication Data
is available

For further information on
Fishing News Books, visit our website:
http://www.blacksci.co.uk/fnb

In memory of my father.
And for my children, wife and mother.

Contents

Preface	xiii
Acknowledgments	xv

1 Overview	**1**
1.1 Rationale	1
1.2 Content	2
1.3 References	3

2 Gene Action I: Qualitative Traits	**4**
2.1 Introduction	4
2.2 Theory	4
2.2.1 Chromosomes, loci, and alleles	4
2.2.2 Dominance	6
2.2.3 Formation of gametes	6
2.3 Practice	9
2.3.1 Ascertaining qualitative inheritance	9
2.4 Illustrative investigations and applications	16
2.4.1 Simple inheritance: a production-related trait	16
2.4.2 Inheritance of color and coloration patterns	17
2.4.3 Albinism	25
2.5 References	30

3 Gene Action II: Inheritance of Quantitative Traits	**33**
3.1 Introduction	33
3.2 Theory	34
3.2.1 Genetic effects and phenotypic variation	34
3.2.2 Average effects and dominance deviations	35
3.2.3 Attributing observed variation to genetic effects	36
3.2.4 Utility of estimates of genetic variation: heritability	43
3.3 Practice	44
3.3.1 Directed mating	44
3.3.2 Identifying or segregating family groups	45
3.3.3 Constraints: analysis and interpretation	50
3.4 Notable investigations and applications	52
3.4.1 Interpreting and applying heritability estimates	52
3.4.2 A case study: *Ictalurus punctatus*	54

		3.4.3	A case study: *Macrobrachium rosenbergii*	56
		3.4.4	A case study: *Procambarus clarkii*	57
		3.4.5	A case study: *Sparus aurata*	59
		3.4.6	Growth, survival, conformation and dressout traits	60
		3.4.7	Disease resistance	63
	3.5	References		64

4 Selection and Realized Heritability   66
    4.1 Introduction   66
    4.2 Theory   67
        4.2.1 Estimating and predicting heritability   67
        4.2.2 Applying selection   70
        4.2.3 Correlated responses   72
        4.2.4 Multi-trait approaches   73
        4.2.5 Complicating and constraining factors   75
        4.2.6 Improving selection efficiency   76
        4.2.7 Using family data   77
    4.3 Practice   78
        4.3.1 Implementation difficulties   78
        4.3.2 Identification options   79
        4.3.3 Lack of response to selection   80
    4.4 Illustrative investigations and applications   80
        4.4.1 Evaluating available strains   80
        4.2.2 Domestication selection   81
        4.4.3 Conflicting results   82
        4.4.4 Correlated responses   83
        4.4.5 Indirect selection through production practices   84
        4.4.6 Indirect measurement   84
        4.4.7 Altering environmental tolerances   85
        4.4.8 Adjusting data for environmental bias   86
        4.4.9 Accounting for differences between sexes   86
        4.4.10 Genotype by environment interactions   86
        4.4.11 Miscellaneous results: finfish   87
        4.4.12 Miscellaneous results: mollusks   88
        4.4.13 Miscellaneous results: crustaceans   89
    4.5 References   90

5 Inbreeding, Crossbreeding and Hybridization   93
    5.1 Introduction   93
    5.2 Theory   94
        5.2.1 Dominance effects and multi-locus traits   94
        5.2.2 Population genetics and dominance effects   95
        5.2.3 Molecular genetics and dominance effects   99
        5.2.4 Utilizing dominance effects for genetic improvement   99
        5.2.5 Alternate goals in hybridization trials   100
    5.3 Practice   100

	5.3.1	Inbreeding impacts	100
	5.3.2	Exploiting heterosis in a production environment	101
	5.3.3	Maternal effects	102
	5.4.3	Combining strain or species attributes	104
	5.3.5	Monosex and sterile hybrids	104
	5.3.6	Combining appropriate broodstock and gametes	104
	5.3.7	Crossbreeding or hybridization in breed formation	105
5.4	Illustrative investigations and applications		107
	5.4.1	A case study: carp and related species	108
	5.4.2	A case study: dominance effects in salmonids	109
	5.4.3	Monosex hybrid stocks	111
	5.4.4	Unexpected results	111
	5.4.5	Results: miscellaneous fish	112
	5.4.6	Hybrid fish for stocking natural waters	115
	5.4.7	Examples: invertebrates	115
5.5	References		116

6 Chromosomal Genetics I: Gynogenesis and Androgenesis 120

6.1	Introduction		120
6.2	Theory		120
	6.2.1	Meiosis and polar bodies	120
	6.2.2	Meiotic gynogenesis	122
	6.2.3	Mitotic gynogenesis	123
	6.2.4	Androgenesis	124
6.3	Practice		125
	6.3.1	Heterozygous versus homozygous gynogenesis	125
	6.3.2	Determining appropriate procedures	127
	6.3.3	Evaluating success	129
	6.3.4	Bivalves	130
	6.3.5	Maternal influences	131
6.4	Illustrative investigations and applications		131
	6.4.1	Gynogenesis	131
	6.4.2	Androgenesis	136
6.5	References		137

7 Chromosomal Genetics II: Polyploidy 141

7.1	Introduction		141
7.2	Theory		142
	7.2.1	Mechanics of inducing polyploidy	142
7.3	Practice		144
	7.3.1	Shock-induced versus interploid triploidy	146
	7.3.2	Bivalve polyploidy	148
7.4	Illustrative investigations and applications		149
	7.4.1	Triploidy in tilapia	149
	7.4.2	Cyprinid results	149

	7.4.3	Salmonid polyploidy	150
	7.4.4	Various other finfish findings	152
	7.4.5	Bivalve studies	154
	7.4.6	Evaluating polyploidy induction	157
7.5	References		158

8 Sex Determination and Control — 162

8.1	Introduction		162
8.2	Theory		163
8.3	Practice		164
	8.3.1	Homogametic monosex stocks	164
	8.3.2	Heterogametic monosex stocks	165
	8.3.3	Minor genetic and environmental influences	165
8.4	Illustrative investigations and applications		166
	8.4.1	A case study: tilapias	166
	8.4.2	Various other finfish	169
	8.4.3	Crustacean sex control and determination	170
8.5	References		171

9 Control and Induction of Maturation and Spawning — 174

9.1	Introduction		174
9.2	Theory		175
	9.2.1	The role of external stimuli	175
	9.2.2	Internal processes	176
	9.2.3	Artificial induction	177
9.3	Practice		178
	9.3.1	Photothermal conditioning and maturation	178
	9.3.2	Hormone-induced spawning	179
	9.3.3	Hormone preparation and injection practices	186
	9.3.4	Injection protocols	189
	9.3.5	Hormone implant technologies	189
	9.3.6	Determining maturational status	192
	9.3.7	Holding and handling broodstock	192
	9.3.8	Obtaining gametes: finfish	193
9.4	Illustrative investigations and applications		198
	9.4.1	Historical development	198
	9.4.2	A case study: the red drum (*Sciaenops ocellatus*)	200
	9.4.3	A case study: the striped bass (*Morone saxatilis*) and its hybrids	202
	9.4.4	Induction with asynchronous ovarian development	206
	9.4.5	Hormone injection of various fish	207
	9.4.6	Crustaceans	211
	9.4.7	Mollusks	212
9.5	References		213

10 Transgenic Aquatic Organisms — 218

10.1	Introduction	218

	10.2	Theory		219
	10.3	Practice		219
		10.3.1	Microinjection	219
		10.3.2	Electroporation	220
		10.3.3	Biolistics	220
		10.3.4	Lipofection	220
		10.3.5	Incorporation and integration	221
	10.4	Illustrative investigations and applications		221
		10.4.1	Transgenesis in Indian catfish	221
		10.4.2	Transgenic tilapia	222
		10.4.3	Atlantic salmon	222
		10.4.4	Carp	223
	10.5	References		224
11	**Genetic Threats to Wild Stocks and Ecosystems**			**225**
	11.1	Introduction		225
	11.2	Theory		225
	11.3	Illustrative investigations		226
	11.4	References		228

Index 229

Preface

The potential for increased production of aquatic species throughout the globe through genetic improvement holds out hope for improved nutrition and a better livelihood for millions of people. The concept for this book arose with this in mind. The purposes envisioned early in its development quickly proved too lofty, however, and as this work took shape it became apparent that, rather than trying to change the world with a single volume, more practical editorial approaches would probably be required.

Genetics is generally a difficult topic to teach, or to learn for that matter, without some occasional simplification. Given the intended audience for this book (non-geneticists with an interest in aquaculture) it became apparent that in order to communicate basic concepts, mechanisms and approaches a degree of simplification would be required to hold readers' interest and attention. As a result, however, this text must serve only as a starting point for the study of genetics in aquaculture.

This book does not present exhaustive methodology for various disciplines of genetic improvement and analysis, but rather strikes some balance between practical application and more technical considerations. Citations have been identified not only for their illustrative value, but also to serve as starting points for those wishing to pursue particular topics in greater depth. Although many of the principles of genetic improvement in aquatic organisms are fairly straightforward, the mechanics and nuances of applying these techniques often take many years to master. The works cited throughout this book reflect the dedicated efforts of many scientists over the years who share my interest in this topic. It is my hope as I finalize this work that it will capture the interest of a new generation of aquaculture geneticists who, collectively, may realize my initial objectives.

Acknowledgments

Many persons have knowingly or unknowingly assisted in the development of this work. I owe a great debt to William R. Wolters, Robert P. Romaire, Arnold M. Saxton and James W. Avault, Jr for their inspiration and encouragement. Just as importantly, however, I must recognize the influence of many of the authors cited in this book, without whom I would not have reached a level of understanding sufficient to attempt this work: Allen, Arai, Avtalion, Beardmore, Becker, Behrends, Bongers, Cherfas, Dobosz, Doyle, El Gamal, Gall, Gjedrem, Gjerde, Gorshkov, Gorshkova, Goryczko, Guo, Falconer, Hershberger, Hickling, Hulata, Hussain, Kirpichnikov, Komaru, Komen, Lerner, Lester, Malison, McAndrew, Meyers, Richter, Rothbard, Rottmann, Shelton, Taniguchi, Tave, Wiegertjes, Wohlfarth, and numerous others. Additionally, my editor has been extremely understanding and professional during the development of this work.

And, finally, I thank Janine, Alexander and Caroline Lutz for their immense patience and loving endurance through the evolution of this manuscript.

Greg Lutz
2001

Chapter 1
Overview

1.1 RATIONALE

Although aquaculture is practiced on a global basis, a tremendous range of technology levels can be found across species, regions, countries and farming operations. Levels of capitalization and access to technical assistance and information are disparate, at best. While many researchers and private corporations are pursuing the potential gains of molecular genetics and other highly technical approaches to the improvement of aquacultured species, the vast majority of aquaculture producers throughout the world have little access to or interest in these approaches.

Most farmed fish, mollusks and crustaceans represent unselected, semi-natural stocks and/or isolated populations suffering from inbreeding, indirect selection and general inferiority to potentially available alternatives. In light of nothing more than the sheer numbers and geography involved, it will probably be many years before the benefits of current molecular biotechnology breakthroughs trickle down to the typical artisanal fish farmer in a developing nation such as Vietnam, Honduras or Nigeria, to name just a few. Nonetheless, real progress can be attained in the interim under these circumstances through more practical methods of genetic improvement. Research has demonstrated repeatedly that practical application of traditional animal and plant breeding methodology can produce substantial gains in productivity in many aquatic species. Additional benefits can often be gained through techniques of chromosomal manipulation entirely unavailable to breeders of traditional terrestrial livestock.

Tremendous gains have been achieved in dairy cattle and poultry production over the past century, in terms of what can be produced on a per animal basis. Havenstein *et al.* (1994) presented compelling data to suggest that almost all of the 300–400% increase in growth rate in modern chickens over the past four decades can be attributed to selection response. The extent to which these gains have found their way to producers and consumers in lesser-developed regions of the world has been somewhat variable. In contrast, virtually every aquaculture industry in every region of the world, including many small-scale artisanal operations, can generate positive gains in productivity over a comparatively small period of time through the application of simple principles of selection and breeding.

The generally high fecundity of many cultured aquatic species allows for much greater selection pressure than can be exercised in most forms of livestock or fowl. Production efficiency, profitability and product attributes such as coloration or conformation are all areas that can be positively impacted through simple, practical genetic applications. Gjedrem (2000) indicated that roughly 65% of the salmon and trout produced in Norway were genetically improved fish resulting from the Norwegian National Breeding Program. The program's cost-benefit ratio was estimated to be 1:15.

In a variety of aquaculture production systems, growth is the production trait most closely correlated with profitability. Although growth is the first trait that comes to mind for most aquaculture species when genetic improvement is mentioned, disease resistance, food conversion efficiency, and many qualitative traits can also be improved through practical approaches. Numerous disease problems have emerged over several decades in a variety of aquacultured species, including shrimp, salmon, channel catfish, sea bream and sea bass, to name but a few. Many of these pathogens were virtually unknown or had never been described prior to major outbreaks in cultured stocks. A wealth of research data suggests that practical genetic improvement can likely be brought to bear to reduce the impacts of these and other, as yet unknown, diseases that may arise in the future.

Similarly, the increasing pressure throughout the globe to reduce environmental impacts of aquaculture serves to emphasize the importance of developing production stocks that convert available food resources more efficiently. Genetic improvement certainly has an important role to play in efforts to further reduce the potential environmental impacts of aquatic animal production.

1.2 CONTENT

The term "gene" presents problems in a book such as this. Depending on the context, gene can refer to a physical site on a chromosome or to one of several specific sets of genetic instructions normally found at that site. Occasionally, other interpretations of this word are also implied. A more thorough treatment of this topic can be found in Chapter 2.

The terms genotype and phenotype will be used extensively in the discussion of many topics in this text, so a clear understanding of these concepts is in order. The physical characteristics of an organism, those measurable or observable traits we can characterize or assign values to, are generally referred to as the organism's phenotype(s). The phenotype, in terms of appearances or performance, reflects to some degree both the environment the organism has been exposed to and the genotype, or genetic composition of the organism itself. Determining the degree to which genotype and phenotype are correlated is a major goal in animal and plant breeding, as will become apparent in Chapters 3 and 4.

The ultimate goal of most investigations into animal and plant genetics is an elusive concept generally referred to as improvement. How one defines improvement, however, depends almost entirely on the species and circumstances in question, but for the purposes of aquaculture genetic improvement generally is equated with higher yields or increased profitability. Lutz (2000) demonstrated the impacts of increased growth on production costs in a variety of tilapia production systems, citing potential reductions in break-even prices of 6% to 14% associated with a 25% reduction in the final growout period. Genetic change in numerous other traits of aquatic species can also result in increased yields and profitability, however, and many will be discussed throughout this book, especially in Chapters 5 through 8.

Control over reproduction is essential to successful genetic improvement in most aquatic species, and various techniques and approaches which address this need are presented in Chapter 9. The extension of these concepts to the development and culture of transgenic aquatic organisms is reviewed in Chapter 10. And, finally, the risks to aquatic habitats and populations posed by all forms of genetically "improved" aquacultured species are considered briefly in Chapter 11.

1.3 REFERENCES

Gjedrem, T. (2000) Genetic improvement of cold-water fish species. *Aquaculture Research*, **31** (Suppl. 1), 25–33.

Havenstein, G., Ferket, P.R., Scheideler, S.E. & Rives, D.V. (1994) Carcass composition and yield of 1991 vs 1957 broilers when fed 'typical' 1957 and 1991 broiler diets. *Poultry Science*, **73**, 1795–1804.

Lutz, C.G. (2000) Production economics and potential competitive dynamics of commercial tilapia culture in the Americas. In: *Tilapia Aquaculture in the Americas*, Vol. 2 (eds B.A. Costa-Pierce & J.E. Rakocy), pp. 119–132. World Aquaculture Society, Baton Rouge.

Chapter 2
Gene Action I: Qualitative Traits

2.1 INTRODUCTION

Qualitative traits are characteristics which can be defined in simple, discontinuous categories. Examples include distinct color variants, variations in fin size or shape, characteristic scale patterns or distinct flesh coloration. Since these traits are typically controlled by only one or two loci, they lend themselves conveniently as illustrations of the basic principles of inheritance as well as the relationships between genotype and phenotype. In keeping with the practical focus of this book, however, portions of the following discussions will be somewhat simplified for the sake of clarity. Those wishing to explore these areas in more depth are encouraged to secure more focused texts on chromosomal and molecular genetics.

2.2 THEORY

2.2.1 Chromosomes, loci, and alleles

In most organisms that concern producers or researchers within the field of aquaculture, chromosomes occur in complete sets and, simultaneously, in pairs. A typical organism possesses two complete sets of chromosomes in the nucleus of each cell in its body (except within gamete-producing regions of the ovaries and testes, as will become apparent below). This condition is referred to as the 2N, or diploid, state. One chromosome set is paternal in origin, the other maternal. With the exception of structurally-distinct sex chromosomes, such as X or Y chromosomes in certain species, these two sets are considered homologous, each with the same distinct number, sizes and shapes of chromosomes. Accordingly, the number of chromosome pairs is the same as the number of distinct types of chromosomes. In turn, a pair of homologous chromosomes, or homologues, are not only the same size and shape but contain the same physical sites associated with producing specific gene products.

Chromosomes themselves consist of complex associations of molecules, but primarily extensive chains of base pairs supported by sugars and phosphorus, which become, in effect, extremely long molecules themselves. This particular type of molecule is, of course, known as DNA, which is short for deoxyribonucleic acid. Within DNA molecules, the base pairs actually form two interlocking chains, or strands, known as a double helix. The helix, in turn, can twist on itself to occupy a comparatively small area. For practical purposes, we need not go into great detail concerning the structure of DNA other than to focus on the linear sequences of base pairs and their relationship to the expression of phenotypes in the aquatic organisms we wish to culture or "improve" upon.

Only four compounds are found in the sequence of base pairs at the heart of the DNA double helix: adenine, cytosine, guanine, and thymine. When researchers attempt to ascertain their order within a particular chromosome, or portion of chromosome, they are generally referred to simply as A, C, G and T. As stated previously, these bases occur in paired states, but each base can pair with only one other. The result is A–T pairs or G–C pairs, and it follows that one strand of a double helix can be interpreted (or constructed) based on the sequence of bases found on a complementary strand.

When chromosomes are replicated in normal cell division, an enzyme known as DNA polymerase separates the two strands of a double helix and uses each strand as a template for construction of a new molecule. This is also the basis for the translation of the genetic code contained in the DNA into the processes of life through a similar molecule referred to as RNA. DNA strands serve as templates for the construction of shorter RNA molecules, which contain corresponding sequences of bases. Through RNA, these instructions are transcribed and transferred from the DNA within the cell nucleus into the cell.

There are thousands of chemical reactions occurring at any point in time within cells of higher organisms. Specific sequences of base pairs on chromosomes provide instructions, through RNA molecules, for the production of amino acids within the cell. The resultant combinations of amino acids in turn lead to the elaboration of proteins and the maintenance of cell (and organism) viability and function. Each amino acid is associated with one or more unique 3-base sequences. Since there are only four bases involved, this allows for a total of 64 (4^3) unique sequences to code for the 20 amino acids. In this way, genetic instructions within a chromosomal segment are transcribed, interpreted and expressed.

Along distinct portions of a specific chromosome, base pairs code for unique gene products. This physical location of instructions on a chromosome is referred to as a locus. The gene product(s) encoded for at any particular locus may pertain to one or many chemical processes within the cells in which they are expressed. A number of distinct forms of instructions may be found at any given locus, however, and these distinct forms of instructions are known as alleles. When two distinct alleles are present at a given locus (one sequence on one chromosome and another on its homologue), one or both may be expressed.

2.2.2 Dominance

This leads to the concept of allelic dominance: either complete or partial dominance. If both sets of instructions are identical for a given locus, the organism in question is said to be homozygous for that particular allele at that locus. This is frequently the case if selection or inbreeding have eliminated all but one possible allele at a given locus (see Chapter 5). If, however, two distinct alleles are present on the two chromosomes including the locus in question, the individual is referred to as heterozygous at that particular locus. Where traits such as color or finnage are involved in aquatic species, many heterozygotes are easily distinguished phenotypically as a reflection of the two distinct alleles present.

In some situations, however, heterozygotes are indistinguishable (phenotypically) from homozygotes. In this situation, the allele for which homozygosity is indistinguishable from heterozygosity is referred to as dominant, and the other allele found within the heterozygote is referred to as recessive. In contrast, if the heterozygote state is clearly distinguishable from either homozygote, the relationship between alleles is referred to as incomplete dominance. Occasionally, several possible alleles may occur at a single locus, resulting in complex patterns of not only dominance, but phenotypic expression as well. To further complicate matters, all degrees of dominance can be found in various traits in aquatic species, and frequently the alleles at two or more distinct loci interact in unexpected ways to influence the phenotypes we observe, as we will see in the following discussions.

The simplest approaches to exploration of the relationships between genetic instructions contained on DNA and the actual phenotype of the organism in question involve those traits that are controlled, or determined, by the alleles present at one, or possibly two, loci. These types of traits are usually observed in discreet, discontinuous categories. In turn, the relative proportions of these categories within a population (or family) can provide insights into the genetic control of the trait. This is possible because the categories can be interpreted, and discerned, in terms of the alleles producing the specific phenotypes we observe, within both parents and offspring. The basis for this approach is the dissolution of pairs of alleles within an organism during the formation of gametes and the subsequent re-establishment of allelic pairing within offspring as gametes combine. This type of inheritance is often referred to as Mendelian, in reference to the early work of Mendel related to the transmission of qualitative traits from one generation to the next.

2.2.3 Formation of gametes

As an organism develops, or survives from day to day for that matter, its cells replicate. Within a cell nucleus, each chromosome is used as a blueprint for the construction of a new, identical chromosome, via the DNA polymerase enzyme mentioned above. As the cell prepares for division through the process of mitosis, for a brief period four sets of chromosomes (4N) are present. Subsequently, two sets (one of paternal origin and one of maternal) are pulled to either side of the cell and the cell's contents are then divided by the cell membrane to form two new, roughly identical, cells (Fig. 2.1).

Conversely, in the formation of gametes, referred to as meiosis, the normal 2N nuclear contents must be transformed into 1N (haploid) eggs or sperm, which can in turn combine to form new diploid (2N) individuals. The initial step in meiosis involves a doubling of the genetic material in each of these two sets of chromosomes, resulting in a temporary 4N condition. However, the two copies of each chromosome remain together during this stage of the process. While all four copies of each distinct type of chromosome are present in primary spermatocytes or primary oocytes, they typically align with each other. This allows segments of homologous DNA strands to break in the same locations and exchange paternal and maternal sequences within specific chromosomal regions. This phenomenon is referred to as crossing over, and it provides for a substantial increase in the variety of gametes, and therefore offspring, an organism can produce.

In the case of sperm production, or spermatogenesis, the original cell contents of the primary spermatocyte, which contains both replicated chromosome sets (4N), are divided to form two secondary spermatocytes, each of which contains one replicated chromosome set (2N). Replicated homologues within each pair are separated: that of maternal origin moves to one secondary spermatocyte while that of

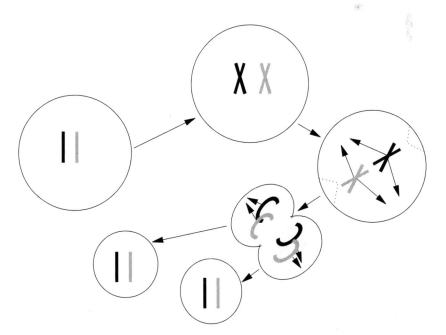

Fig. 2.1 The process of normal cell division, or mitosis, somewhat simplified. Note chromosomes replicate and daughter cells each contain a complete complement.

paternal origin moves to the other. For each type of chromosome present, however, this process occurs independently, resulting in combinations of maternally and paternally derived chromosomes within each secondary spermatocyte. A similar process occurs within secondary spermatocytes, with replicated portions of distinct chromosomes finally separating and subsequently being divided among four resultant sperm cells, each with a complete 1N set of chromosomes (Fig. 2.2).

In this way, if more than one allele is present at a given locus within an organism, that is, if the organism is heterozygous at that locus, two of the four sperm cells produced from a primary spermatocyte will contain copies of one allele and two will contain copies of the other. Perhaps even more importantly, unless two loci are in close physical proximity along the same chromosomal segment, their alleles will find their ways into primary and secondary spermatocytes and oocytes independently.

Egg formation, or oogenesis, follows the same general pattern as spermatogenesis, with eggs ultimately containing 1N chromosome sets including maternally and paternally derived genetic material (Fig. 2.3). In egg formation, however, cell contents are usually conserved throughout the process of oogenesis and are ultimately utilized by only one of the four haploid sets of chromosomes. While this solitary 1N genome combines with a sperm cell in the formation of a new organism, the other three chromosome sets are lost, in what are referred to as the first and second polar bodies. The first polar body is formed at the same time as the secondary oocyte, and contains one replicated but undivided set of chromosomes (2N). The second polar body and the egg proper each contain 1N of genetic material from the final separation of the replicated chromosome set present in the secondary

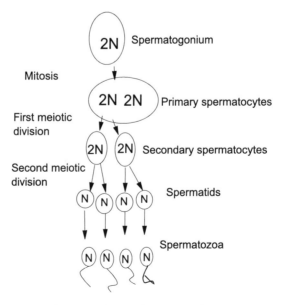

Fig. 2.2 The process of meiosis during formation of spermatozoa.

Qualitative Traits 9

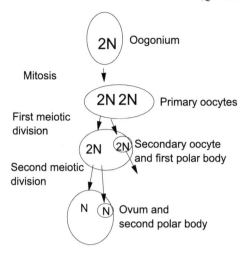

Fig. 2.3 The process of meiosis ultimately yielding an egg pronucleus and the first and second polar bodies.

oocyte. Typically, the second polar body is not lost until the egg is fertilized, at which point the sperm activates a number of internal processes within the egg.

Within the egg, in addition to the 1N of nuclear genetic material, a number of mitochondria (intra-cellular, double-membrane bounded organelles) are typically present. Found in all eukaryotic cells, mitochondria are essential in the production of adenosine triphosphate (ATP). Mitochondria contain their own DNA and the necessary mechanisms for duplicating it and themselves. Accordingly, mitochondrial inheritance is entirely maternal, and may play an important role in maternal influences on complex traits, as we will discuss in later chapters.

2.3 PRACTICE

2.3.1 Ascertaining qualitative inheritance

When a novel qualitative trait is discovered, say, an attractive color variant, the first step to take is to isolate the individual and ultimately allow it to spawn. If one or more novel individuals of each sex are available, they can be crossed directly to examine the inheritance of the trait in question. Often, however, only a single variant is available. In these situations, the novel individual should be spawned with one or more normal-colored, or wild-type, individuals (preferably unrelated to the color variant or each other). Complications can arise of course if the organism possessing a desirable novel phenotype happens to be female and the normal reproductive biology of the species limits it to one mate per spawn. Additionally, male aggressiveness during courtship and spawning (as might occur in bettas, tilapia or many other aquatic species) can exacerbate these problems and lead to risk of losing the unique genetic code from which the phenotype may originally have been expressed.

One potential benefit of spawning a novel variant to a number of normal individuals is simply to produce large numbers of offspring that carry the allele(s) responsible for the trait in question (if, in fact, the variation is not the result of some environmental factor). Although the first generation of offspring from these matings (often referred to as the F_1 generation) may occasionally provide some insight into the genetic control of the trait in question, it is often necessary to allow these F_1 offspring to mate among themselves in order to shed light on the numbers and types of loci and alleles involved. This approach also reduces to some degree the inevitable inbreeding that arises when trying to increase numbers of a rare phenotype.

More often than not, the F_1 offspring produced by crossing a novel individual with one or more normal individuals present normal phenotypes. Even if this is not the case, the F_2 generation produced by crossing F_1 individuals will usually provide easily-characterized phenotypic categories in proportions that can provide insights into the alleles producing the specific phenotypes we observe, within both parents and offspring.

Relating phenotype to genotype

Once we know the genetic control of a particular novel trait (or combination of traits) we can easily predict breeding outcomes, but working backward to understand mechanisms of control requires a try-it-and-see methodology. Assume a minnow grower discovers a fish in his stock that is golden in color, as opposed to the normal gray-green phenotype of the species he produces. Conveniently, for this discussion, the minnow happens to be a male. The grower places the male with a number of females and collects the eggs. The F_1 offspring produced are all gray-green in color, as might be expected.

Subsequently, the minnow grower diligently raises the F_1 fish to breeding age and allows them to spawn among themselves, isolated from the rest of his breeding stock. As the F_2 generation grows, it becomes apparent that some golden individuals are present. Sampling reveals there are roughly three gray-green individuals for every golden one. These results would suggest that the golden phenotype results from a single locus being homozygous for a recessive allele (Fig. 2.4). The golden allele can be considered recessive because those individuals that carry both golden and normal (gray-green) alleles are phenotypically normal.

Consider another example, still using the minnow operation described above. The minnow producer this time is presented by one of his workers with a novel fish that is pink and has elongated fins. Again, the grower spawns this novel fish and the F_1 offspring produced are all gray-green in color with normal finnage. If the subsequent F_2 generation included roughly three gray-green normal-finned individuals for every one that was pink with long fins and no other phenotype classes

Gg female × Gg male

Eggs \ Sperm	G	g
G	GG	Gg
g	Gg	gg

GG = Gray-green (1/4)
G**g** = Gray-green (2/4)
gg = golden (1/4)

Fig. 2.4 Progeny classes from mating two Gg heterozygotes (see text).

were apparent, the results would suggest that the pink coloration and long fin phenotypes both resulted from a single controlling locus being homozygous for a recessive allele (Fig. 2.5).

More likely in the real world, however, would be the following scenario. The F_1 offspring produced would all be gray-green in color with normal finnage. Sampling in the subsequent generation would reveal a number of phenotypes: gray-green with normal fins, gray-green with long fins, pink with normal fins, and a few pink individuals with long fins. After repeated sampling, these phenotypes would occur in a ratio of roughly 9:3:3:1. These results would suggest that the color and finnage traits were independently controlled by two distinct loci, and within each locus the novel alleles, for pink coloration or long fins, were recessive to those producing the normal phenotype (Fig. 2.6).

If the alleles in question were not completely dominant, such that intermediate phenotypes were associated with the heterozygous state for each locus, things would be somewhat easier to interpret, if not to tabulate. Consider the case where a minnow that was heterozygous for the pink allele was, phenotypically, a color reminiscent of rust. Similarly, we will imagine that minnows heterozygous for the allele producing long fins had intermediate fin lengths. The expected results would include an even greater variety of phenotypes, with every combination of coloration and finnage.

Specifically, the color phenotypes of gray-green, rust and pink would occur in a 1:2:1 ratio independent of finnage, and the finnage

Nn female × Nn male

Eggs \ Sperm	N	n
N	NN	Nn
n	Nn	nn

NN = Normal (Gray-green, normal fins) (1/4)
N**n** = Normal (Gray-green, normal fins) (2/4)
nn = Pink with long fins (1/4)

Fig. 2.5 Progeny classes from mating two Nn heterozygotes (see text).

Pp Ff female × Pp Ff male

Eggs\Sperm	P F	P f	p F	p f
P F	PP FF	PP Ff	Pp FF	Pp Ff
P f	PP Ff	PP ff	Pp Ff	Pp ff
p F	Pp FF	Pp Ff	pp FF	pp Ff
p f	Pp Ff	Pp ff	pp Ff	pp ff

PP FF, Pp FF, PP Ff & Pp Ff = Gray, normal fin (9/16)
PP ff & Pp ff = Gray, long fin (3/16)
pp FF & pp Ff = Pink, normal fin (3/16)
pp ff = Pink, long fin (1/16)

Fig. 2.6 Progeny classes from mating two double heterozygotes, Pp Ff, complete dominance.

phenotypes of normal, intermediate and long would occur in the same ratio, respectively, independent of coloration. These ratios relate to homozygous and heterozygous states for the alleles in question. When overlaid upon each other, as it were, we see that for every four rust-colored minnows with intermediate fin length (double heterozygotes) there should be one gray-green normal finned fish and one pink long-finned fish (double homozygotes). The overall ratio of expected phenotypes, with normal, intermediate and long fins within gray-green, rust and pink colored fish, respectively, would be $1:2:1:2:4:2:1:2:1$ (Fig. 2.7).

Clearly, the proportions of distinct phenotypes are most illustrative when heterozygotes can be distinguished. By looking for associations with specific ratios of phenotypes, it is often possible to develop a hypothesis for genetic control of one or more traits, which can subsequently be tested through examination of progeny phenotypes from mating specific individuals, such as backcrossing. Certain situations, however, can present more cryptic results, especially when more than one locus is involved in the expression of a single particular phenotypic

Eggs\Sperm	P F	P f	p F	p f
P F	PP FF Gray, Norm.	PP Ff Gray, Med.	Pp FF Rust, Norm.	Pp Ff Rust, Med.
P f	PP Ff Gray, Med.	PP ff Gray, Long	Pp Ff Rust, Med.	Pp ff Rust, Long
p F	Pp FF Rust, Norm.	Pp Ff Rust, Med.	pp FF Pink, Norm.	pp Ff Pink, Med.
p f	Pp Ff Rust, Med.	Pp ff Rust, Long	pp Ff Pink, Med.	pp ff Pink, Long

Gray, Normal: (1/16)	Rust, Normal: (2/16)	Pink, Normal: (1/16)
Gray, Medium: (2/16)	Rust, Medium: (4/16)	Pink, Medium: (2/16)
Gray, Long: (1/16)	Rust, Long: (2/16)	Pink, Long: (1/16)

Gray : Rust : Pink = 1:2:1   Normal : Medium : Long = 1:2:1

Fig. 2.7 Progeny classes from mating two double heterozygotes, no dominance.

variant. In fact, it is often possible for multiple genotypes to produce identical phenotypes. Consider the example below of the "pearl" phenotype in the Nile tilapia (*Oreochromis niloticus*).

A case study: "pearl" Nile tilapia

Over the years, color inheritance in tilapia species and hybrids has been addressed by various authors (Tave 1991; Lutz 1997), but some recent tilapia breeding work may represent one of the more complex examples of qualitative inheritance in a food-fish species (Lutz 1999). In the mid-1990s, a tilapia hatchery in Louisiana reported some unusual Nile tilapia (*Oreochromis niloticus*) fingerlings. These fish were noticeably white, close to the color of milk. Between 20 and 30 fish exhibited this white coloration out of roughly 15 000 fingerlings. The hatchery owner-operator indicated that other groups of these light-colored fish had been noticed over the previous few months, but no effort had been made to set them aside.

Fry were typically harvested from as many as 12–15 tanks of brood fish in one day, so it was unclear in which tank(s) these fish were being produced. Upon further investigation, it became apparent that white fingerlings were originating from one particular tank of brood fish, which held 4–5 males and a number of females – all with perfectly normal coloration for Nile tilapia. At the time, it was not possible to determine which specific combination or combinations of broodstock within the tank were producing these unusual offspring.

Many markets for live tilapia in North America place extra value on silvery or light-colored fish, so the market potential for these white fish among growers already buying from the hatchery was immediately apparent. In order to ascertain the genetic control, if any, over this trait, these light-colored Nile tilapia fingerlings were set aside for growout on site. In previous reports in the literature, the only mention of such light-colored tilapia usually involved crosses of certain red hybrids. Although the stock of Nile tilapia in question had been electrophoretically certified, a thorough examination of characteristics such as scale and fin-ray counts was conducted to determine if they were actually Nile tilapia. As the fingerlings grew, they appeared to possess all the meristic characteristics of Nile tilapia, and developed accessory coloration (black fin edges, pink highlights, etc.) typical of the species (Fig. 2.8). Additionally, their coloration became much more intense than the uniform off-white exhibited earlier in life. Based on their opalescence, the phenotype was designated as pearl.

Upon reaching sexual maturity, a number of the pearl-colored fish were crossed with each other. Although it initially appeared that the offspring displayed a 50% white: 50% normal color ratio, Chi-square tests on a number of broods from these fish indicated the ratio was statistically closer to 9:7 (white to normal). This implied more than one locus, probably two, was involved in the control of this trait, and development of a line of pearl-colored fish would be somewhat difficult.

The 9:7 ratio of offspring from these "pearls," combined with the normal coloration of their parents suggested a theory for the control of

14 *Practical Genetics for Aquaculture*

Fig. 2.8 The pearl strain of Nile tilapia (*Oreochromis niloticus*) generally exhibits the accessory pigments associated with mood and behavior but lacks most of the skin pigments normally found in this species. (Photo courtesy I. Bubeck.)

this trait, based on the phenotypic ratios of 9:3:3:1 developed in the examples above. For simplicity's sake, two loci, designated as A and B, were hypothesized. Each locus would of course have two alleles: for either normal coloration (A and B), or "recessive" alleles (a and b). The a and b alleles in combination, and only in combination, would result in pearl coloration. Consider the theory that each of the normal-colored parents of the original pearl fish carried "recessive" alleles (alternative alleles might be a better description) at only one of the two loci involved. One parent would be represented as AaBB, and the other as AABb. Either recessive allele, when present alone, would not alter the normal coloration of the fish, but if one copy of each a and b were present, the result would be pearl coloration.

If this were true, what would have been the genotypes of the original pearl fingerlings? Remember that the gametes contributed by the parent fish would include all possible combinations of the alleles from each locus (Fig. 2.9). Under this hypothesis, the AaBb fish produced would exhibit the pearl phenotype. To carry this hypothesis one step farther, the expected outcome when these original pearl fish were mated with each other was determined based on the assumption of an AaBb genotype (Fig. 2.10). The pearl-colored individuals, shown in Fig. 2.10 as underlined genotypes, would make up 9/16 of the offspring from such a cross if the hypothesis were valid, assuming the presence of at least one copy of a and at least one copy of b would result in a pearl phenotype.

AaBB female × AABb male

Eggs \ Sperm	AB	aB
AB	AABB	AaBB
Ab	AABb	AaBb

AABB, AaBB, AABb = Normal
AaBb = Pearl (1/4)

Fig. 2.9 Hypothesized allele transmission of broodstock yielding first generation pearl-colored Nile tilapia.

AaBb × AaBb

Eggs \ Sperm	AB	Ab	aB	ab
AB	AABB	AABb	AaBB	AaBb
Ab	AABb	AAbb	AaBb	Aabb
aB	AaBB	AaBb	aaBB	aaBb
ab	AaBb	Aabb	aaBb	aabb

AABB = Normal (1/16)   AABb = Normal (2/16)   AAbb = Normal (1/16)
AaBB = Normal (2/16)   AaBb = Pearl (4/16)   Aabb = Pearl (2/16)
aaBB = Normal (1/16)   aaBb = Pearl (2/16)   aabb = Pearl (1/16)

Fig. 2.10 Progeny classes produced from mating two first generation pearl Nile tilapia.

An important aspect of this explanation for the inheritance pattern of the pearl phenotype was that one out of every nine pearl offspring produced from the original pearl broodstock would be a double homozygote (aabb) that could produce all-pearl offspring even when crossed with a normal (AABB) individual (Fig. 2.11). Subsequent progeny testing of individuals from the second generation of pearls identified both males and females that produced all-pearl offspring when mated to other strains of Nile tilapia. Another outcome of this pattern of control over the pearl phenotype is that several genotypes result in

AABB female × aabb male

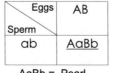

Eggs \ Sperm	AB
ab	AaBb

AaBb = Pearl

Fig. 2.11 Expected progeny when crossing a double-homozygous pearl Nile tilapia.

AaBB female × AABb male

Eggs / Sperm	AB	aB
AB	AABB	AaBB
Ab	AABb	AaBb

AABB, AaBB, AABb = Normal
AaBb = Pearl (1/4)

Fig. 2.12 Outcome of crossing normal-colored AaBB and AABb Nile tilapia.

aaBb female × aaBb male

Eggs / Sperm	aB	ab
aB	aaBB	aaBb
ab	aaBb	aabb

aaBb, aabb = Pearl (3/4)
aaBB = Normal (1/4)

Fig. 2.13 Outcome of crossing partially heterozygous pearl-colored Nile tilapia.

indistinguishable pearl phenotypes: AaBb, aaBb, aabb, and Aabb. As a result, mating second generation pearls among themselves resulted in several ratios of pearl- to normal-colored offspring (Figs 2.12 and 2.13).

Since commercial breeding of large numbers of fry usually requires large numbers (10–40) of broodfish to be housed in groups, it becomes difficult to simply segregate pearl-colored broodstock for production of fry without producing some normal-colored fish. With multiple genotypes expressed as identical pearl phenotypes, progeny testing has proven difficult other than to identify those (aabb) individuals that consistently produce all-pearl offspring. Commercial production of pearl heterozygotes will probably require breeding programs similar to those of scaled/leather/mirror carp (Fig. 2.14).

2.4 ILLUSTRATIVE INVESTIGATIONS AND APPLICATIONS

2.4.1 Simple inheritance: a production-related trait

Forward *et al.* (1995) reported results that suggest that susceptibility of brook charr, *Salvelinus fontinalis* to the haemoflagellate *Cryptobia salmositica* is controlled by a single locus with dominant (R) and recessive (r) alleles. Milt from four susceptible male charr and four resistant male

Fig. 2.14 The mirror form of common carp (*Cyprinus carpio*) occasionally appears in normally-scaled lines and can be produced predictably by crossing male and female leather carp. The mirror phenotype results from a homozygous recessive condition and generally breeds true. (Photo courtesy J.W. Avault, Jr.)

charr were used to fertilize eggs from two susceptible females. All progenies of one resistant male were resistant to infection, although their mothers were susceptible. The male was presumed to be RR and the females rr. The other three resistant males produced progeny in a ratio of roughly 1:1 susceptible to resistant, when mated to susceptible females. These resistant males were presumed to be Rr. The four susceptible males produced susceptible offspring when mated with susceptible females, and all were presumed rr.

2.4.2 Inheritance of color and coloration patterns

Ornamental fish
Frequently, development of novel strains involves substantial alteration of phenotypes (Fig. 2.15). Numerous examples of qualitative inheritance can be found in ornamental fish (Fig. 2.16). It would be beyond the scope of this work to address them all, but several examples may be in order to illustrate the types of traits and methodologies involved. Fernando and Phang (1990) provided insight into the genetic control of color patterns in three varieties of the domesticated guppy (*Poecilia reticulata*). Virtually all strains of guppies are sexually dimorphic, especially in terms of coloration. The authors crossed wild-type guppies with individuals from the tuxedo, blonde tuxedo, and red snakeskin varieties, and subsequently crossed the resultant offspring to examine phenotypic classes in the F_2 generation. They determined

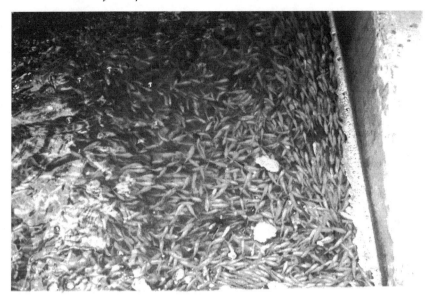

Fig. 2.15 In the southern US, the "rosy-red" race of fathead minnows have been developed to a point where recommended cultural practices differ notably from those applied to wild-type strains. (Photo by the author.)

Fig. 2.16 Inheritance of qualitative characters in many ornamental aquatic species, once ascertained, has often been considered proprietary information and thus not widely published. (Photo by the author.)

that dominant sex-linked alleles at X- and Y-linked loci (see Chapter 8 for X-Y sex determination) produced the black caudal peduncle (Bcp) and red tails (Rdt) associated with Tuxedo and Blonde Tuxedo strains.

In the blonde strain, however, males were heterozygous for the black peduncle allele, and an autosomal (non-sex related) recessive allele not only resulted in the blonde background color but also interacted with the Bcp and Rdt alleles, diluting the black and red "tuxedo" coloration in males. They further determined that alleles at two Y-linked loci produced snakeskin patterns in the bodies (Ssb) and tails (Sst) of red snakeskin males. The Rdt and Sst alleles also appeared to interact to produce black tail spots in these males. Investigations such as this are tedious at best, but often justified when development of proprietary ornamental varieties is a realistic goal.

Perhaps one of the most widely-known examples of the inheritance of complex coloration patterns can be found in ornamental strains of *Cyprinus carpio*: koi carp. Originally named nishikigoi by their breeders in Niigata, Japan, almost two centuries ago, these fish have been adopted by hobbyists and breeders throughout the world. Koi exhibit a wide range of color variation, pattern variation, and combinations thereof. So complex are these variations that investigations of their inheritance seem daunting to most breeders and geneticists. As a result, limited formal analyses have been attempted to date (Katasonov 1978; Wohlfarth & Rothbard 1991).

Gomelsky *et al.* (1996) indicated that red-white color complexes and black spots or patches appear to be independently inherited in koi (Fig. 2.17). Based on gynogenetic trials, they concluded that a number of loci are probably involved in red-white coloration patterns in this fish. Many other traits are displayed in koi, however, such as shades

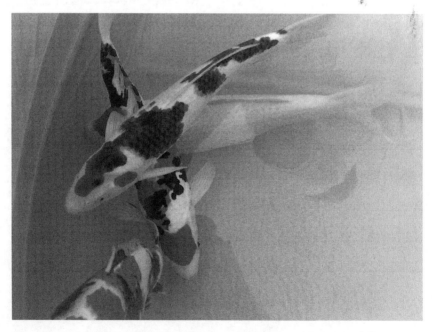

Fig. 2.17 Findings suggest that variable red-white coloration and patterns of deposition of black pigment in koi carp (*Cyprinus carpio*) are under separate genetic control. (Photo courtesy B. Nichols.)

of golden yellow, unusual scale patterns, and occasional presence of metallic aspects in scale reflectance. Each may exhibit its own patterns of inheritance. Valuation of koi is based on specific coloration and pattern combinations; more valuable fish generally display less common coloration or color patterns, and these forms generally do not breed true (Fig. 2.18). Wide variation in coloration and color patterns is typically observed even when broodstock of similar coloration and pattern are crossed with each other (Fig. 2.19). Nonetheless, many breeders contend that the frequency of certain phenotypes may increase noticeably based on certain combinations of broodstock. As a result of high phenotypic variation and difficulty in predicting the types of coloration and color patterns a given cross may produce, fish are often individually inspected at an early age to select those with the most valuable phenotypes.

A review: color inheritance in tilapia
Tave *et al.* (1989) described the simple autosomal mode of inheritance of a gold phenotype in *Oreochromis mossambicus*. Gold individuals were shown to be homozygous for the recessive allele (g) at the locus in question, while wild-type (normal colored) individuals were homozygous for the normal (G) allele. Interestingly, heterozygotes (Gg individuals) displayed a "bronze" skin color. The authors suggested the ability to produce true-breeding all gold populations could be a valuable tool for genetic studies with *O. mossambicus*.

Many other distinct color variants have been described in tilapia populations. Tave (1991) reviewed the simple inheritance of several ti-

Fig. 2.18 Fingerling koi carp. Even when phenotypically similar individuals are spawned, a wide variety of colors and patterns typically result. (Photo by the author.)

Fig. 2.19 Sorting through offspring of champion koi can be painstaking work. On this occasion, only 11 of 722 individuals were considered promising enough to meet or exceed their parents' qualities. (Photo by the author.)

lapia color variants. Three examples came from work at the University of Stirling in Scotland with the Lake Manzala stock of *Oreochromis niloticus*: "blond," "syrup" and "red". The blond (a faded khaki-like appearance) and syrup (orange-yellow) colors were determined to result from recessive alleles at single autosomal (unrelated to sex) loci, each with complete dominance for the wild-type (normal) coloration. A third color variant in this stock, red, also resulted from a single autosomal gene, but the mutant (red) allele was dominant to the wild type. Selection against any carrier of such a mutant allele would probably serve to eliminate it almost as quickly as it arose in a natural population, since both homozygotes and heterozygotes exhibit the red phenotype (Tave *et al.* 1991).

With the exception of koi breeding, the topic of color inheritance in tilapia is probably unrivaled as an illustration of how complicated "simple" inheritance can be in some instances. While certain accounts of color variants in tilapia may seem unrelated, and unimportant at that, an examination of one comparatively unspectacular phenotype a number of years ago can shed some light on the complexity and interrelatedness of patterns of phenotypic expression in different populations. Mires (1988), working at Kibbutz Ein Hamifrantz in Israel, reported on patterns of inheritance of a light-colored phenotype of *Oreochromis niloticus* in a strain from Uganda. The light-colored fish all had a clear peritoneum, and instead of the dark dorso-lateral stripes typical of the species, displayed only various intensities of light gray on the back and tail.

When these individuals were mated randomly within their own stock, all the first generation progeny displayed the wild-type (normal) coloration with distinguishable striping. Inbreeding between these light-colored individuals, however, produced mainly light pink individuals, with smaller numbers of bright red, orange and a few "albino" offspring. Whether these bright red *O. niloticus* were analogous to those described above in the Lake Manzala strain is doubtful, based on patterns of inheritance. They do, however, exhibit a pattern of heredity similar to "red" tilapia hybrids described in some other studies, as we will see below.

Anyone familiar with the global tilapia industry can appreciate the importance of the emergence of numerous colorful phenotypes, mostly resulting from the appearance of a few red individuals in another tilapia species almost 30 years ago in Taiwan. Ho Kuo, an accomplished researcher from the Taiwan Fisheries Research Institute, spent years developing the commercial potential of a gift of about 100 red *O. mossambicus* variants. Over the course of two decades, he determined this inheritance of the recessive trait in *O. mossambicus* and developed hybrids with *O. niloticus*, noting the expression of black markings in the F_1 hybrids and describing selection efforts to virtually eliminate these markings and develop stable red and white strains. His presentation at the Second International Symposium on Tilapia in Aquaculture (Kuo 1988) is a valuable reference for anyone studying the history of red tilapia.

While volumes have been written concerning the production, genetics and economic and marketing merits of "red" tilapia, the topic of "albino" tilapia, as referred to by both Mires (1988) and Kuo (1988), is problematic. These phenotypes are intimately related in terms of inheritance and expression, and most accounts of the "albino" or true white phenotype are associated with studies of the inheritance of red coloration. Many instances of truly white tilapia have been noted, and these phenotypes are not, apparently, analogous to the silvery hybrid-derived strains marketed in North America. These white tilapia typically display normal pigment deposition within the eyes, resulting in gray-black pupils, as opposed to the shiny pink appearance found in true albinos of other fish species, including the tilapiine genera *Pseudotropheus* and *Haplochromis*. The observed mode of inheritance (or lack thereof) for this black-eyed phenotype in a variety of tilapia stocks (and related species such as *Pseudotropheus zebra*) also suggests it may not represent any form of albinism in the traditional sense.

In a population derived from *O. niloticus* by *O. mossambicus* hybrids, McGinty (1983) proposed simply that normally colored individuals and white or pink fish were homozygous and red individuals were heterozygous. Examining F_2 progeny, produced by allowing black-flecked red-gold hybrids to mate randomly, Behrends et al. (1982) described tilapia phenotypes including wild type, reddish brown with darkly pigmented dorsal area, reddish gold with black patches, pink and reddish orange mottled with white, continuous pink with no black markings, and continuous off-white. Interestingly, some phenotypes were predominately female (wild type, reddish brown, pink-orange with white mottling) while others were predominately male (reddish gold with patches and continuous pink).

In a more complex analysis, possibly as a result of a different base population, Galman *et al.* (1988) outlined the production of a number of easily characterized color variants from random and directed breeding of Philippine red tilapia. This stock resulted from crossing of *O. niloticus* with red tilapia introduced from Singapore in 1978. When "red-orange gold with black spots" founder stocks were allowed to spawn, a number of phenotypes were expressed in the offspring, including red gold-orange without and with spots (designated by the authors as phenotypes I and II), uniform pink without and with spots (phenotypes III and IV), "albino with black eyes" (phenotype V) and uniform gray (phenotype VI).

Halstrom (1984) related progress in reduction of black markings on red tilapia in Taiwan through selective breeding. The actual physiology of color expression in the pure red mossambicus, Philippine red tilapias (hybrids) and normal-colored tilapia was described by Avtalion and Reich (1989). Crosses within phenotype I (gold-orange without spots) produced I, III, IV and V phenotypes, with no normal-colored fry. Crosses within phenotype III (uniform pink without spots) produced only the I, III, and V phenotypes, also with no normal-colored progeny and without the spotted pink individuals reported from the other cross.

When type III males were crossed with type I, II and IV females, type VI (normally gray) individuals comprised 25% ± 7.7%, 23% ± 6.9% and 20% ± 13.4%, respectively, of the resulting progeny. All non-gray phenotypes were present in various proportions in the progeny of each cross, with the exception of type IV progeny from the III male by I female cross. In short, a number of parental phenotypic combinations produced the "albino with black eyes" type V phenotype, although it was apparently not abundant in any instance. Galman *et al.* (1988) referred to type V as a "lethal trait" phenotype, possibly implying the white by white cross failed to produce viable offspring. Similar observations of the general lack of vigor of type V phenotypes have been noted by this author and others (L.L. Behrends personal communication, 1987).

The white phenotype, however, can apparently play an important role in the mass production of red tilapia in some populations. At the Ein Yahav station in Israel, Koren *et al.* (1994) examined the colored inheritance of two types of crosses yielding all-red progeny. Male white tilapia, segregates from random-mating Philippine red tilapia and apparently analogous to the type V phenotype described by Galman *et al.* (1988), when mated to normal colored females from the same stock, to female *O. aureus*, or to female *O. niloticus*, consistently produced all-red progeny. In the same study, male "dominant red" phenotypes of *O. niloticus*, when mated to normal *O. niloticus* females, also produced all-red progeny.

When Hulata et al. (1995) bred "red" female *O. niloticus* to *O. niloticus* males, all-red progeny were obtained. Based on this pattern of inheritance these fish were probably exhibiting the same dominant red mutation described above for Manzala *O. niloticus* and seen in the Koren et al. (1994) study. In contrast to Koren et al.'s success with white tilapia males, however, in the same study the authors reported that *O. aureus* females mated to white segregates (again from Philippine red tilapia) produced progeny in a ratio of 35% red to 65% bronze-colored individuals.

When pink, gold, or orange tilapia are referred to as "red", especially when research results are being translated from Chinese to English or Spanish or Hebrew, confusion can result and often does. The "dominant red" phenotype of *O. niloticus* utilized in the Koren et al. (1994) and Hulata et al. (1995) studies is probably under somewhat different genetic control than the hybrid phenotypes used by Galman et al. (1988). Phenotypic expression in hybrid populations should not be assumed or expected to follow patterns exhibited by pure species.

Behrends and Smitherman (1984) and Behrends et al. (1988) documented the development of a cold-tolerant strain of red tilapia attained through hybridization and recurrent backcrossing with a cold-tolerant line of *O. aureus*. In this breeding scheme, F_1 "hybrids" and backcross progeny were composed of approximately 50% red- and 50% normal-colored progeny. Based on this pattern of inheritance, the authors proposed a simple heterozygous determination of red coloration. Similar explanation for the inheritance of red coloration was proposed by Hussain (1994) in his study of Thai and Egyptian strains of red tilapia developed from *O. niloticus* and red *O. mossambicus*.

Further work with the populations described by Behrends et al. (1988) and similar studies with *O. niloticus* and its red hybrid led the authors to propose that two independent "genes", the Mm locus determining melanistic pigmentation and the Rr locus determining red coloration, resulted in the production of normal, pink, red, or white individuals (L.L. Behrends personal communication, 1987). Similarly, in his previous study of Taiwanese red tilapia (based on hybrids of the red *O. mossambicus* described by Ho Kuo with *O. urolepis hornorum*) Halstrom (1984) had suggested the genetic control of coloration depended on two loci, R and M, with recessive epistasis (interaction), where the expression of the genotype at the M locus depends on the alleles at the R locus.

Some interesting details emerged from the Behrends et al. (1988) study, including a preponderance of males (84%) in the red phenotype prior to selection. Although the red phenotype displayed higher mortality than their normal-colored counterparts (37% vs. 15%), when early growth advantages were eliminated red phenotypes were 7% heavier on average. Red phenotypes exhibited a greater selection response, and selection for size resulted in a correlated response in percent males, for both red- and normal-colored phenotypes. This positive correlation between growth and sex (maleness) cannot be dismissed as a random effect, sampling bias, or probably even as differential survival among sexes within treatments, for Hulata *et al.* (1986) and Lester *et al.* (1989) also reported relationships between growth rate and progeny sex ratios in populations of *O. niloticus*.

El Gamal *et al.* (1988) observed that although crosses between red- and normal-colored hybrids of *O. aureus* and between similar *O. niloticus* hybrids produced red- and normal-colored embryos in a 1:1 ratio, hatchability of red embryos was significantly reduced. While viability during a hatchery and sex-reversal period (see Chapter 8) was essentially the same for both phenotypes, survival of the red siblings was significantly lower during subsequent growout in concrete tanks. The fact that all progeny were sex-reversed prevented an objective evaluation of the influence of sex on survival in red hybrids. At any rate, in a commercial setting this type of protracted mortality for red phenotypes could translate into significant reductions in the efficiency of food conversion, as individuals that have been fed and grown for a substantial portion of the production cycle are reduced to little more than fertilizer.

A number of other red tilapia varieties have arisen or been developed over the years, some of which display much better survival than the strains examined by El Gammal *et al.* (1988). Although single-species vs. hybrid ancestry appears to be a key factor in predicting the inheritance of body color in tilapia stocks, substantial work has yet to be done to clarify and hopefully unify our understanding in this area of aquaculture genetics.

2.4.3 Albinism

Albinism is, perhaps, the most familiar type of recessive trait with simple inheritance recognized by aquaculturists the world over. Numerous albino ornamental fish can be obtained in most parts of the world. Albinos occur in a variety of cultured species, including salmonids, cichlids, cyprinids and characins, to name just a few groups. They are often put to good use in studies of gynogenesis and androgenesis, where their developing eggs can provide visible confirmation of the presence or absence of genetic contributions from normal-colored individuals (see Chapter 6).

Albinos, in the strict sense of the word, reflect a genetic condition that does not allow the production of melanin, perhaps one of the most common pigments in most animals, but many other white- or light-colored genetically-based variants can be found in a variety of species and these are sometimes referred to as albinos. By the 1950s, much work had already been accomplished in the characterization and description of albinism in many ornamental fish species. An interesting review of albinism in ornamental fish can be found in Goldstein (1996). While older accounts of albinos in aquatic species commercially cultured for food are few and scattered, more reports have surfaced in recent years that focus on cultured species or their close relatives. An interesting case study involving the Indian carp *Catla catla* was presented by Bhowmick *et al.* (1986). Through induced spawning (see Chapter 9), a stock of catla was developed over a period of years from a few initial fingerlings.

Albinism in commercial species can result in more marketable products under some circumstances, but if predation is a problem in culture ponds, tanks or raceways albinos may suffer disproportionately due to their higher visibility. Nelson (1959) reported all-albino progeny were produced when albino channel catfish (*Ictalurus punctatus*) were crossed with each other. The albino parents, from a domesticated strain, had appeared in spawns produced by normal-colored fish. Nelson indicated these albinos appeared to be equal in "ruggedness" to their normal-colored counterparts. He speculated that the normal turbidity of catfish culture ponds might offset potential sensitivities to sunlight that plague some albino fish. Additionally, these fish were referred to as "golden" channel catfish, due to the effect of pigments other than melanin, which were either acquired through the diet or elaborated within the fish themselves.

Prather (1961) also reported similar growth rates between albino and normal colored channel catfish (Fig. 2.20). Survival, however, was somewhat lower for albinos, and Prather speculated this was the result of increased vulnerability to predation due to the high visibility of these fish when feeding. Several later studies in Arkansas, Alabama and Georgia also indicated no substantial differences in the performance or hardiness of albino channel catfish, and no consistent trend in survival differences. In contrast, in their examinations of albinism in the grass carp *Ctenopharyngodon idella*, Rothbard and Wohlfarth (1993) indicated that when albinos and wild-type carp were grown together

Fig. 2.20 Channel catfish (*Ictalurus punctatus*) feeding on floating pellets. Note high visibility of albino individuals, although most of the population is normally colored. (Courtesy James W. Avault, Jr.)

in cages, albino fish were significantly smaller. The fact that products from specific loci can be expected to influence more than one trait could account for impacts of albinism on performance traits that might, at first consideration, seem independent of coloration. This is not, however, always the case for color variants (Fig. 2.21). Inbreeding effects (see Chapter 5) are often implicated when albino stocks are developed from small numbers of individuals.

Albino individuals in other types of catfish have also been described. Many are familiar with the albino armored catfish (*Corydoras* sp.) widely available in the aquarium trade throughout the world. Some other examples include the African sharptooth catfish *Clarias gariepinus* (Hoffman *et al.* 1995), the wels *Silurus glanis* (Dingerkus *et al.* 1991), the pimelodid *Rhamdella minuta* (Sazima & Pombal 1986), black bullheads *Ictalurus melas* (Hicks 1978) and the marine catfish *Tachysurus tenuispinis* (Baragi *et al.* 1976). Many other catfish, mostly from the ornamental trade, exhibit albinism commonly.

More than one locus may be involved in albinism, either singly or in combination. In goldfish, two loci are involved simultaneously in the expression of albinism. Similarly, in the simultaneous hermaphrodite pulmonate snail *Physa heterostropha*, Dillon and Wethington (1994) used two distinct loci, each of which could result in albinism, to study mating systems and inbreeding depression. In the closely related frogs *Rana nigromaculata* and *R. brevipoda*, Nishioka *et al.* (1987) described five distinct loci that could result in albinism.

Fig. 2.21 In addition to blue and white (albinistic) forms, a rare yellow-orange phenotype of the red swamp crawfish *Procambarus clarkii* (seen here beside a normally colored individual) has been found on several occasions in both Louisiana and Spain. These color variants exhibit no noticeable reduction in growth or fitness. (Photo by the author.)

In rainbow trout and steelhead, both races of the same species (*Oncorhynchus mykiss*), at least two distinct loci can produce albinism, individually. Thorgaard *et al.* (1995) reported that although albino rainbow trout from four domesticated strains and one Idaho steelhead strain are homozygous recessive at the same locus, an albino steelhead strain from Washington state is apparently mutant at a distinctly different locus, inasmuch as crossing this albino strain with the others results in pigmented fry. Albinism is common in other salmonids as well (Fig. 2.22). Dauble *et al.* (1978) reported on an albino chinook salmon (*Oncorhynchus tshawytscha*) taken from a wild population in the Columbia River.

Dobosz *et al.* (1999) reported on unrelated studies of albinism in rainbow trout. In this species, a homozygous recessive allele at one particular locus results in yellow, rather than normal, coloration. Among these yellow fish, a dominant allele at a second locus controls the palomino coloration and normal black eye color. A recessive allele at this second locus, when homozygous, results in pink eyes and a lack of pigment deposition within the yellow trout line.

Nakatani (1999) reported on the occurrence of an "albino" red swamp crawfish, *Procambarus clarkii*. The animal was all white, lacking melanin or any of the other pigments normally found in the exoskeleton, but the eyes were normally pigmented. Nakatani established, by mating this "albino" to a normal-colored crawfish and subsequently mating the resulting normal-colored offspring with each other, that this trait was the result of a single, recessive allele in the homozygous state. This same type of "albino" crayfish, lacking pigment throughout the body but with normally pigmented eyes, was reported many years

Fig. 2.22 Albino salmonids have been widely cultured in a number of countries. (Courtesy James W. Avault, Jr.)

ago in a related species, *P. alleni*, by Taborsky (1982). Again, those crossings that were possible suggested the albino condition was a result of a homozygous recessive allele.

Black (1975) reported on studies of the inheritance of blue coloration in the commercially important crayfish *Procambarus zonangulus* (formerly *P. acutus acutus*). The blue coloration was determined to be controlled by a recessive allele that resulted in conjugation of the normal pigment, astaxanthin, with protein. This same mutation is common in *P. clarkii*. Many years ago, I had similar results to those reported by Nakatani when spawning a female "albino" *P. clarkii* with a blue-pigmented variant of this species. The eggs were white rather than the normal brownish-black color for this species, reflecting the female's inability to produce pigments normally present in oocytes, but as the resultant juveniles began to grow and produce their own pigment the resulting offspring were normal colored.

The fact that "albino" crayfish exhibit normally colored eyes would suggest that a separate locus and/or biochemical pathway is involved in the elaboration of eye pigment (Fig. 2.23). This hypothesis is supported by the sporadic occurrence of "white-eyed" or "golden-eyed" individuals in *Procambarus clarkii*. This condition is inherited as a simple recessive trait (Black & Huner 1980), and results from the absence of distal and retinular screening pigments, as well as the fact that crayfish hemolymph is clear rather than red in coloration. Interestingly, Krouse

Fig. 2.23 Although extremely rare, individuals of several decapod species have been reported with a form of albinism that excludes eye pigments, which are apparently under separate genetic control. (Photo by the author.)

(1981) described two "albino" rock crabs, *Cancer irroratus*, from Penobscot Bay, Maine, with the same appearance: a very white body but normally colored, black eyes. This suggests the possibility that many decapods may possess the same distinct pathways for eye and other body pigments.

2.5 REFERENCES

Avtalion, R.R. & Reich, L. (1989) Chromatophore inheritance in red tilapias. *Israeli Journal of Aquaculture/Bamidgeh*, **41** (Suppl. 3), 98–104.

Baragi, V.M., Yaragal, R.B. & James, P.S.B.R. (1976) On an albino of the marine catfish, *Tachysurus tenuispinis* (Day). *Matsya*, **2**, 82–83.

Behrends, L.L. & Smitherman, R.O (1984) Development of a cold-tolerant population of red tilapia through introgressive hybridization. *Journal of World Aquaculture Society*, **14**, 172–178.

Behrends, L.L., Kingsley, J.B. & Price, A.H. (1988) Bidirectional backcross selection for body weight in a red tilapia. In: *The Second International Symposium on Tilapia in Aquaculture* (eds R.S.V. Pullin, T. Bhukaswas, K. Tonuthai & J.L. Maclean), pp. 125–133. International Center for Living Aquatic Resources Management, Manila, Philippines.

Behrends, L.L., Nelson, R.G., Smitherman, R.O. & Tone, N.M. (1982) Breeding and culture of the red-gold color phase of tilapia. *Journal of the World Mariculture Society*, **13**, 210–220.

Bhowmick, R.M., Kowtal, G.V., Jana, R.K. & Gupta, S.D. (1986) Observations on hypophysation and rearing of albino catla, *Catla catla* (Hamilton). *Indian Journal of Animal Sciences*, **56** (Suppl. 4), 482–484.

Black, J.B. (1975) Inheritance of the blue color mutation in the crawfish *Procambarus acutus acutus* (Girard). *Proceedings of the Louisiana Academy of Sciences*, **38**, 25–27.

Black, J.B. & Huner, J.V. (1980) Genetics of the red swamp crawfish, *Procambarus clarkii. Proceedings of the World Mariculture Society*, **11**, 535–543.

Dauble, D.D., Gray, R.H., Hanf, R.W. & Poston, T.M. (1978) Occurrence of a wild albino chinook salmon (*Onchorhynchus tshawytscha*) in the Columbia River. *Northwest Science*, **52** (Suppl. 2), 108–109.

Dillon, R.T. & Wethington, A.R. (1994) Inheritance at five loci in the freshwater snail, *Physa heterostropha. Biochemical Genetics*, **32** (Suppl. 3–4), 75–82.

Dingerkus, G., Seret, B. & Guilbert, E. (1991) The first albino wels, *Silurus glanis* Linnaeus, 1758, from France, with a review of albinism in catfishes (Teleostei: Siluriformes). *Cybium*, **15** (Suppl. 3), 185–188.

Dobosz, S., Goryczko, K., Kohlmann, K. & Korwin-Kossakowski, M. (1999) The yellow color inheritance of rainbow trout. *Journal of Heredity*, **90** (Suppl. 2), 312–314.

El Gamal, A.A., Smitherman, R.O. & Behrends, L.L. (1988) Viability of red and normal-colored *Oreochromis aureus* and *O. niloticus* hybrids. In: *The Second International Symposium on Tilapia in Aquaculture* (eds R.S.V. Pullin, T. Bhukaswas, K. Tonuthai & J.L. Maclean), pp. 153–157. International Center for Living Aquatic Resources Management, Manila, Philippines.

Fernando, A.A. & Phang, V.P.E. (1990) Color pattern inheritance in three domesticated varieties of guppy, *Poecilia reticulata. Aquaculture*, **85**, 320.

Forward, G.M., Ferguson, M.M. & Woo, P.K.T. (1995) Susceptibility of brook charr, *Salvelinus fontinalis* to the pathogenic haemoflagellate, *Cryptobia salmositica*, and the inheritance of innate resistance by progenies of resistant fish. *Parasitology,* **111** (Suppl. 3), 337–345.

Galman, O.R., Moreau, J., Hulata, G. & Avtalion, R.R. (1988) Breeding characteristics and growth performance of Philippine red tilapia. In: *The Second International Symposium on Tilapia in Aquaculture* (eds R.S.V. Pullin, T. Bhukaswas, K. Tonuthai & J.L. Maclean), pp. 169–175. International Center for Living Aquatic Resources Management, Manila, Philippines.

Goldstein, R.J. (1996) Albino fishes. *Aquarist and Pondkeeper,* **61**, 22–25.

Gomelsky, B., Cherfas, N.B., Ben-Dom, N. & Hulata, G. (1996) Color inheritance in ornamental (koi) carp (*Cyprinus carpio* L.) inferred from color variability in normal and gynogenetic progenies. *Israeli Journal of Aquaculture/Bamidgeh,* **48** (Suppl. 4), 219–230.

Halstrom, M.L. (1984) *Genetic studies of a commercial strain of red tilapia.* M.S. Thesis, Auburn University.

Hicks, D.C. (1978) A population of albino black bullheads, *Ictalurus melas. Copeia,* **1978** (Suppl. 1), 184–185.

Hoffman, L.C., Prinsloo, J.F., Theron, J. & Casey, N.H. (1995) A chemical comparison between the golden and normal colored strains of the African sharptooth catfish, *Clarias gariepinus* (Burchell 1822). *Journal of Applied Ichthyology/Zeitschrift fur angewandte Ichthyologie,* **11** (Suppl. 1–2), 71–85.

Hulata, G., Karplus, I. & Harpaz, S. (1995) Evaluation of some red tilapia strains for aquaculture: Growth and colour segregation in hybrid progeny. *Aquaculture Research,* **26** (Suppl. 10), 765–771.

Hulata, G., Wohlfarth, G.W. & Halevy, A. (1986) Mass selection for growth rate in the Nile tilapia (*Oreochromis niloticus*). *Aquaculture,* **57**, 177–184.

Hussain, M.G. (1994) Genetics of body color inheritance in Thai and Egyptian red tilapia strains. *Asian Fisheries Science,* **7** (Suppl. 4), 215–224.

Katasonov, V.Ya. (1978) Color in hybrids of common and ornamental (Japanese) carp III. Inheritance of blue and orange types. *Soviet Genetics,* **14**, 1522–1528.

Koren, A., Pruginin, Y. & Hulata, G. (1994) Evaluation of some red tilapia strains for aquaculture. 1. Color inheritance. *Israeli Journal of Aquaculture/Bamidgeh,* **46** (Suppl. 1), 9–12.

Krouse, J.S. (1981) Occurrence of albino rock crabs, *Cancer irroratus* Say (Decapoda, Brachyura) in Penobscot Bay, Maine, U.S.A. *Crustaceana,* **40** (Suppl. 2), 219–220.

Kuo, H. (1988) Progress in genetic improvement of red hybrid tilapia in Taiwan. In: *The Second International Symposium on Tilapia in Aquaculture* (eds R.S.V. Pullin, T. Bhukaswas, K. Tonuthai & J.L. Maclean), pp. 219–221. International Center for Living Aquatic Resources Management, Manila, Philippines.

Lester, L.J., Lawson, K.S., Abella, T.A. & Palada, M.S. (1989) Estimated heritability of sex ratio and sexual dimorphism in tilapia. *Aquaculture and Fisheries Management,* **20**, 369–380.

Lutz, C.G. (1997) What do you get when you cross...? Part 2: Colors in Tilapia. *Aquaculture Magazine,* **23** (Suppl. 3), 90–96.

Lutz, C.G. (1999) The Nile pearl: an unusual color morph in *Oreochromis niloticus. Aquaculture Magazine,* **25** (Suppl. 6), 74–77.

McGinty, A.S. (1983) Genetics and Breeding: Puerto Rico. In: *Southern Region Cooperative Research Project S-168. Annual Report 1982. Warmwater Aquaculture* (ed. R.R. Stickney), p. 13. Alabama Agricultural Experiment Station, Auburn University.

Mires, D. (1988) The inheritance of black pigmentation in two African strains of *Oreochromis niloticus*. In: *The Second International Symposium on Tilapia in Aquaculture* (eds R.S.V. Pullin, T. Bhukaswas, K. Tonuthai & J.L. Maclean), pp. 237–241.

International Center for Living Aquatic Resources Management, Manila, Philippines.

Nakatani, I. (1999) An albino of the crayfish *Procambarus clarkii* (Decapoda: Cambaridae) and its offspring. *Journal of Crustacean Biology*, **19** (Suppl. 2), 380–383.

Nelson, B.A. (1959) Progress report on golden catfish. *Proceedings of the Annual Conference of the Southeastern Association of Game and Fish Commissioners*, **12**, 75–78.

Nishioka, M, Ohtani, H. & Sumida, M. (1987) Chromosomes and the sites of five albino gene loci in the *Rana nigromaculata* group. *Scientific Reports of the Laboratory for Amphibian Biology, Hiroshima*, **9**, 1–52.

Prather, E.E. (1961) A comparison of production of albino and normal channel catfish. *Proceedings of the Annual Conference of the Southeastern Association of Game and Fish Commissioners*, **15**, 302–303.

Rothbard, S. & Wohlfarth, G.W. (1993) Inheritance of albinism in the grass carp, *Ctenopharyngodon idella*. *Aquaculture*, **115** (Suppl. 1–2), 13–17.

Sazima, I. & Pombal, J.P. (1986) An albino of *Rhamdella minuta*, with notes on behaviour (Osteichthyes, Pimelodidae). *Revista Brasileira de Biologia*, **46** (Suppl. 2), 377–381.

Taborsky, P. (1982) The white prince. *Tropical Fish Hobbyist*, **08/82**, 8–13.

Tave, D. (1991) Genetics of body color in tilapia. *Aquaculture Magazine*, **17** (Suppl. 2), 76–79.

Tave, D., Rezk, M. & Smitherman, R.O. (1989) Genetics of body color in *Tilapia mossambica*. *Journal of the World Aquaculture Society*, **20** (Suppl. 4), 214–222.

Tave, D., Rezk, M. & Smitherman, R.O. (1991) Effect of body colour of *Oreochromis mossambicus* (Peters) on predation by largemouth bass, *Micropterus salmoides* (Lacepede). *Aquaculture and Fisheries Management*, **22**, 149–153.

Thorgaard, G.H., Spruell, P., Wheeler, P.A. *et al.* (1995) Incidence of albinos as a monitor for induced triploidy in rainbow trout. *Aquaculture*, **137** (Suppl. 1–4), 121–130.

Wohlfarth, G.W. & Rothbard, S. (1991) Preliminary investigations on color inheritance in Japanese ornamental carp (*nishiki-goi*). *Israeli Journal of Aquaculture/Bamidgeh*, **43**, 62–68.

Chapter 3
Gene Action II: Inheritance of Quantitative Traits

3.1 INTRODUCTION

Many traits pertaining to the commercial performance of aquaculture species are described as metric, continuous, or quantitative characters. Metric characters involve the end result of complex physiological functions that are nonetheless measurable as simple, distinct variables. The continuous nature of metric characters is often manifested in frequencies approximating normal distributions. Typical aquaculture-related examples include growth rate, feed conversion efficiency, and tolerance of extremes of temperature, salinity or dissolved oxygen. Quantitative characters can include traits measured in the form of dimensions or counts, such as fecundity, dressout percentage, or various morphometric dimensions. Simply put, variation in metric characters leads to questions like "Why is that prawn so much larger than the rest?" or "Why is that carp so broad-bodied?" or, in situations many aquaculturists can relate to, "Why are these few fingerlings still alive after the rest have all died?"

As discussed in the previous chapter, non-continuous mechanisms govern many inheritance patterns observed in aquatic species, especially qualitative characters such as coloration. Nonetheless, seemingly qualitative traits often have an underlying continuous basis. Limitations in the way these traits are expressed can require the division of the population into distinct, discontinuous classes. In the case of reproductive success, for example, although data for any given individual may be interpreted as "yes it spawned" or "no it didn't" during a given spawning season, underlying variation within the population may actually have a normal distribution with some break point, or "threshold," separating the two outcomes.

Since several to many loci (as well as environmental variation) are usually involved in the expression of a continuous trait, direct patterns of inheritance are usually obscure. It is rarely practical or even possible to determine exactly how many loci are involved in the phenotypic expression of such traits. What is possible, however, is estimation of the types of genetic effects that exercise the most influence over a trait and formulation of improvement programs to make the best use of these sources of variation. Ascertaining the genetic basis of observed variation and determining methods for improving a trait usually require examination of variation among various types of relatives, often with the assistance of statistical methods. The statistical treatment of variation, or variance, is the basis for much of the theory and practice relating to improvement of metric characters.

3.2 THEORY

3.2.1 Genetic effects and phenotypic variation

Genetic control of metric traits manifests itself primarily in two familiar animal- and plant-breeding phenomena: (1) resemblance between relatives and (2) heterosis (often referred to as hybrid vigor) or the alternative, known as inbreeding depression. These phenomena are based on additive genetic effects and dominance genetic effects, respectively. When formulating the most promising approach to genetic improvement of any trait expressed by a particular species, the influences of additive and dominance effects will determine the importance of selection, crossbreeding, or a combination of these practices in a breeding program.

Additive effects are directly related to the heritability of various traits and thus are the basis for selection, where using the best-performing individuals as broodstock results in an overall improvement in the average performance of the offspring. Additive effects are the basis of an individual's breeding value, which can be interpreted as the expected performance of its offspring when it is mated at random within a population. Here we touch on the most important aspect of additive genetic effects: their manifestation across generations. The observed variation, or variance, of these breeding values is, accordingly, termed additive genetic variance.

Dominance effects, in contrast, are the result of specific combinations of alleles and, accordingly, combinations of individuals, strains or species that contribute those alleles. This fundamental influence of parental combinations relates directly to the manifestation of dominance effects through heterosis and/or inbreeding depression. Note that the combinations of alleles, the very foundation of dominance effects, are dissolved between generations only to reform on a random basis in each new generation.

Two additional sources of variation that occasionally come into play in aquaculture genetics are epistasis and maternal effects. Epistasis refers to interactions among genetic effects. In reality, most breeding designs and facilities available to aquaculturists are inadequate to evaluate epistasis with much precision. Fortunately, it rarely influences phenotypic variation to such a degree as to require its quantification.

Maternal effects, in contrast to epistasis, frequently contribute tangibly to the phenotypic variation we observe in aquaculture species. These influences often have a genetic basis, and may reflect characteristics such as egg quality or size, care of eggs or fry (as in tilapia or crayfish, for example), or more subtle tendencies such as synchrony of ovulation and spawning. Maternal effects are also often observed in

Inheritance of Quantitative Traits

the application of gynogenesis (Chapter 6), polyploidy (Chapter 7) and hybridization (Chapter 5) in aquatic species.

3.2.2 Average effects and dominance deviations

To proceed with this discussion we must first examine how additive and dominance effects influence phenotypic variation with a simplistic example: at a single locus. Some examples of color inheritance in tilapia (Tave 1991) discussed in the previous chapter will serve to illustrate these basic concepts. When we consider a single-locus trait, it is possible in many instances to determine an individual's genotype by simple observation:

Species:	*Oreochromis mossambicus*		
Trait:	gold/bronze/normal (gray-black) coloration		
Phenotype:	Gold	Bronze	Normal
	−a	0,d	+a
Genotype:	gg	Gg	GG

This situation can be described as a lack of dominance, or incomplete dominance depending on how we score the Bronze phenotype, since the heterozygote Gg individuals can be easily distinguished phenotypically from their GG counterparts. When examining the additive attributes of alleles within a single locus, the values of −a and +a are typically assigned to the two homozygous genotypes (GG and gg, for example). Additionally, a value of 0 is calculated at an intermediate point and, finally, some relative value of d is assigned to the heterozygote (Gg in this case). In reality, d can fall anywhere along the scale from −a to +a. This exercise need not be simply arbitrary. In this instance, one might generate actual mathematical values for these three genotypes based on pigment deposition or light reflectance.

In contrast to the gold/bronze/normal coloration example above in Mozambique tilapia, a heterozygote may often be phenotypically indistinguishable from a homozygote for whichever of the alleles is dominant (as previously considered in Chapter 2). This type of interaction between alleles is referred to as complete dominance:

Species:	*Oreochromis niloticus*		
Trait:	blond/normal (gray) coloration		
Phenotype:	Blonde		Normal
	−a	0	d,+a
Genotype:	blbl		Blbl,BlBl

Again, for this and many similar examples, numerical data based on pigment deposition or reflectance could be generated to mathematically derive values for −a, and a (=d).

An important concept to retain at this point is that as gametes are formed, pairs of alleles are separated through meiosis which ultimately form new genotypes with alleles contributed by other individuals.

The average deviation, from the population mean, of offspring which received a specific allele from one parent combined at random with corresponding alleles at prevailing frequencies within the population is referred to as the average effect of that allele. This average effect can alternately be conceptualized as the result of substituting the allele in question for a representative sample of alleles for that locus from the population as a whole. In a gold/bronze/normal Mozambique tilapia population, the g allele will consistently result in lighter-colored offspring and the G allele in darker-colored offspring. In their directional nature, these results represent additive genetic effects. As stated previously, the additive influence of an allele on the trait in question can be expected to persist from generation to generation.

While these examples neglect to involve allele frequencies for simplicity's sake, the inclusion of prevailing frequencies in our understanding of average effects can have some interesting impacts within a population. Consider the genotypes at the Nile tilapia locus (BlBl, Blbl, blbl) for which dominance is complete, when a recessive allele (bl) is substituted at random. If the recessive allele occurs at a very low frequency to begin with, most of the individuals in the population will be BlBl, and a substitution to produce Blbl will produce little or no change in phenotypic value. The average effect of bl will be small. If the bl allele is more common, however, many of these substitutions will result in conversion of Blbl individuals to blbl, producing a significant change in the population mean and a much larger average effect. The allele's influence on the phenotype is unchanged, but the quantification of its average effect becomes, to a large degree, a reflection of the allele's frequency within the population. Note that the average effect of an allele is not only additive, based on the values of –a and +a, but also persists directionally from generation to generation. This is the key to a number of concepts developed in this and later chapters.

One convenient way to illustrate dominance effects in these examples would be as those genetic effects not accounted for by the sums of average effects within pairs of alleles. This can be seen most easily in the blond/normal Nile tilapia example. In this example the shifting of the value d from 0 to +a represents a dominance deviation. On average, a bl allele does result in a lighter-colored individual, representing an additive effect. Phenotypically, however, Blbl individuals are indistinguishable from BlBl fish. Such interactions between specific alleles which modify, negate, or supersede the sum of their average effects are the result of dominance effects.

3.2.3 Attributing observed variation to genetic effects

When making the jump from simple qualitative traits controlled by one or two loci to metric characters involving several to many loci, we must abandon the idea of mathematically determining additive or dominance genetic effects on a locus-by-locus basis and simply attempt to estimate their overall impact, across all loci involved, on the observed variation of the trait in question. Here we encounter another fundamental concept in aquaculture genetics. When looking to the future for most aquaculture stocks, and the improvements we might wish to make on a practical basis, genetic effects are not nearly as important as the variation attributable to them. This concept is reflected in the types of real-world questions listed in the introduction (page 33), pertaining to commonly observed variation in growth, body conformation or environmental tolerance.

To efficiently assess individual genetic potential in most aquaculture species, we are forced to work with easily measured values. Clearly, the phenotype is the most quantifiable physical expression of most continuous characters we might wish to examine. Unfortunately it typically includes not only various genetic influences, each with its own pattern of variation, but also environmental influences (and, occasionally, genotype-by-environment interaction). In most statistical analyses of variation within cultivated populations, environmental influences are assumed to have a mean of 0, at least within a given generation. The fortunate result of this assumption is the phenotypic mean for a character should approximate the genotypic mean if the sample is large enough to be representative. Values will range around this mean due to a number of sources of variation, or variance.

To determine the relative importance of various sources of genetic variation for a given trait one must not only separate random environmental variation from genetic variation but must further attribute genetic variation to additive effects, dominance effects and occasionally other genetic influences such as epistasis or maternal effects. By examining how the genetic portion of the observed variance is partitioned among different types of relatives, such as full siblings (progeny with both parents in common), half siblings (progeny with only a common father or mother) or unrelated individuals, estimates of the portions of phenotypic variance accounted for by additive, dominance and other genetic effects can be calculated.

This statistical partitioning of distinct genetic variances is made possible by their very nature: additive variance being based on the resemblance among relatives due to directional influences retained by alleles as they are passed from one generation to the next, and dominance variance resulting from interactions of specific combinations of alleles originating from specific parents. When tools such as computers and statistical software are available, the concept of covariance allows application of analyses such as regression or analysis of variance to phenotypic data collected from production stocks or experimental animals. The components of the phenotypic variation we observe, both genetic and environmental, can then be associated with covariances in amounts and proportions directly attributable to the types of relationships between individuals and the manner in which they are raised.

Since the genetic phenomenon of resemblance between relatives is associated with additive genetic effects, the greater the additive portion of genetic variance within a population for a particular trait the greater the expected degree of resemblance, or covariance, for that trait between related individuals. In related individuals, alleles with additive effects that are identical by descent will produce the same phenotypic influences. Similarly, covariances resulting from dominance effects are based on the combination of alleles producing a dominance effect at a particular locus being identical by descent in two or more individuals. Such covariance can occur only when individuals are related to each other on both sides of their pedigrees, as is the case for full-siblings. Unfortunately, being raised under similar environmental conditions can also result in similarities among individuals, related or otherwise, and analyses must be designed to take this tendency into account.

As we examine the genetic and environmental effects that contribute to phenotypic variance, some basic relationships should be defined:

$$V_P = V_G + V_E$$

and

$$V_G = V_A + V_D + V_I$$

where:

V_P = phenotypic (observed) variance,
V_G = genetic variance, composed of
V_A = additive variance
V_D = dominance variance and
V_I = interaction variances, or epistasis,
and V_E = environmental variance.

Epistasis refers to variation resulting from the interaction of alleles at different loci. There are numerous permutations of this type of variance, but in most instances its contribution to the total observed variation does not justify the complex mating designs and analyses required to quantify it. Some degree of epistasis, however, may be phenotypically associated with additive variance, and result in upward bias when this component of variation is being estimated. As a rule, the more complex the interaction, the less the overall expected impact on phenotypic variation.

Additionally, the term V_M can be used to represent maternal effects in some experimental designs, but in most practical analyses these effects can represent confounding genetic or environmental factors depending on the circumstances and traits in question. Examples of genetically based maternal effects might include heritable brooding behaviors or cytoplasmic inheritance. Examples of environmental maternal effects include spawning or nesting locations, and other non-heritable factors. When these distinct types of influences can be distinguished, the terms V_{Mg} or V_{Me} can be used. If egg size or quality are determined to be statistically involved in maternal effects, questions inevitably arise as to the relationship of these variables to nutrition, husbandry practices, genetic effects, or a combination of causes.

Determining efficient approaches to genetic improvement through trial and error has taken many years for some aquatic species, but single-generation analyses can often allow valuable estimates of the similarities, or covariances, between different types of relatives. This is possible because additive effects are associated with resemblance between relatives not only across generations but also within them. For any given estimate of overall phenotypic variance within a population, greater similarity within groups implies greater differences between groups. Depending on whether and to what extent we can group individual animals into full- or half-sib families, the resemblance between relatives within these groups can be contrasted to the differences between groups.

Considering the objective to partition sources of phenotypic variation within a population, it should come as no surprise that experimental designs for single-generation investigations are structured as common statistical analyses of variance (ANOVAs). Parental contributions become effects in the analysis. In these types of analyses, female broodstock have historically been referred to as dams and males as sires. Depending on the types of matings which can be directed and the facilities available, investigations can take on the form of nested designs, factorials, block designs, or other permutations of familiar ANOVAs. Each of a number of males can be mated exclusively to its own separate group of several females (Fig. 3.1), or combinations of males and females can be included in multiple matings (when species' biology allows) to provide multiple progeny groups related in various ways (by sires, dams, or both). Since a review of the statistical theory on which these analyses are based is beyond the scope of this discussion, readers are encouraged to review them in texts on experimental design and analysis.

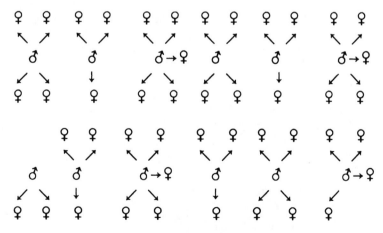

Fig. 3.1 A schematic of a nested mating design for estimation of genetic variances. Note that the number of dams per sire need not be equal to perform a meaningful analysis.

Once the mechanics of producing various mating arrangements have been worked out for the species in question, the next step is to develop an experimental design and analysis that will yield meaningful results. A key requirement is to be able to identify the parentage of any particular offspring, either through markers or segregation in separate ponds, cages or tanks. A nested approach to variance analysis is often the most practical means of meeting all these requirements, especially when full-sib groups can be sub-divided into two or more replicate environments (Table 3.1).

This nested approach has been widely used because the reproductive biology of many aquatic species allows the mating of one male to several females with relative ease over a fairly short period of time. The availability of full- and half-sib groups in the same analysis is also of great benefit, since comparing the covariances for each of these types of relatives provides some indication not only of the contributions of additive variance but also those of dominance and maternal variance, although these effects are somewhat confounded. Solving for the variance components from the ANOVA allows one to proceed with their direct interpretation as numerical estimates of components of genetic variance (Table 3.2). The covariance of paternal half-sibs in a nested design relates strictly to additive effects. The only genetic influence they share directly arises from the average effects of alleles received from their sire. Variation among full-sib progeny groups contains a compo-

Table 3.1 Example of ANOVA for paternal half-sib and full-sib progeny groups and derivation of variance components.

Source of variation	Composition of mean square*
Between sires	$\sigma^2_w + k_1\sigma^2_D + dk_2\sigma^2_S$
Dams (within sires)	$\sigma^2_w + k_3\sigma^2_D$
Between progeny	σ^2_w

*d = number of dams, k values are calculated from the data
σ^2_S = component of variance attributable to sires
σ^2_D = component of variance attributable to dams
σ^2_w = component of variance observed within progeny

Table 3.2 Genetic and environmental interpretation of variance components from a nested mating design.

Component	Genetic effects
Sires (σ^2_S)	$= \frac{1}{4} V_A$
Dams (σ^2_D)	$= \frac{1}{4} V_A + \frac{1}{4} V_D + V_M + V_{Ec}$
Progenies (σ^2_w)	$= \frac{1}{2} V_A + \frac{3}{4} V_D + V_{Ew}$ (overall environment)

Where V_A = additive genetic variance, V_D = dominance genetic variance, V_M = maternal variance, V_{Ec} = variance from common environment, and V_{Ew} = other environmental variance.

nent of dominance variance, since individuals within a full-sib group represent combinations of gametes, and therefore alleles, from the same two individuals.

Maternal effects, however, can also contribute to the covariance within full-sib groups, since all progeny share the same dam. Maternal effects play a major role in most aquaculture species through influences on egg size, egg quality, fertility and larval survival, and through mitochondrial inheritance. For example, Danzmann and Ferguson (1995) reported that significant differences in body size were attributable to specific mitochondrial DNA haplotypes within each of two distinct strains of rainbow trout (*Oncorhynchus mykiss*). When maternal half-sibs can be produced, the resulting factorial analysis can allow for much closer estimation of dominance and maternal effects. While communal rearing of family groups is desirable to provide a common environment, practical means of identifying individual progeny are often unavailable, necessitating separate rearing units for separate family groups (Fig. 3.2). While molecular genetic markers are being developed for a number of aquatic species, their practical application on a farm-level basis is not yet realistic. This situation may change in the future, however. Ideally, all full-sib groups can be subdivided into two or more rearing units (tanks, ponds, net-pens) to account statistically for that component of environmental variance (Fig. 3.3).

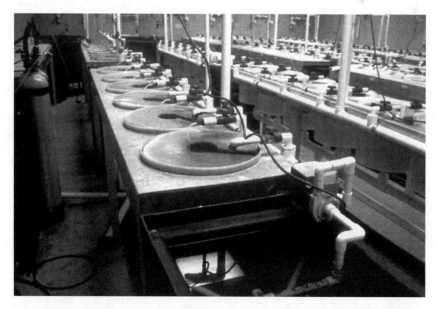

Fig. 3.2 Inability to adequately identify sib groups, especially during early life history, often results in the need to segregate families in individual tanks, cages or pens. One approach to minimize confounding environmental variance involves the use of recirculating systems to provide similar water quality for all sib groups. (Photo by the author.)

Fig. 3.3 When separate pools or ponds must be used to segregate family groups, random or uncontrolled environmental variation may require the use of two or more replicate units per family group. This, in turn, may reduce the number of groups that can be included within an analysis. (Photo by the author.)

Clearly, a simple nested design will not provide the data required to estimate all genetic effects that may influence a trait, but with sufficient numbers of observations (at least 30 per family) it will provide a suitable estimate of additive variance in most cases (Table 3.3). The advantage of a nested mating design lies in its comparative ease to execute and in its use of half-sib families, eliminating many potential sources of bias in the estimation of additive variance. In some situations, variance components may include interaction effects of interest to breeders and statisticians, such as additive by additive or additive by dominance epistasis. Estimation of such effects, however, requires not only large numbers of observations but also complex mating designs and rearing

Table 3.3 Estimation of primary genetic effects based on variance components from a nested mating design.

Values	Effects estimated
$4(\sigma^2_S)$	$= V_A$
$4(\sigma^2_D)$	$= V_A + V_D + 4V_M + 4V_{Ec}$
$2(\sigma^2_S + \sigma^2_D)$	$= V_A + \frac{1}{2} V_D + 2V_M + 2V_{Ec}$

Where V_A = additive genetic variance, V_D = dominance genetic variance, V_M = maternal variance, V_{Ec} = variance from common environment, and V_{Ew} = other environmental variance.

facilities. Such efforts are rarely warranted in terms of the overall contributions of these interactions to total genetic variance. Accordingly, for practical purposes they will not be further addressed in this discussion. Becker (1984) provides an excellent review of the derivation of genetic effects from a variety of mating designs.

3.2.4 Utility of estimates of genetic variation: heritability

Once the observed variation within a population has been partitioned, the results can be directly applied to the formulation of a plan for genetic improvement. A general term used to describe the contribution of additive variance to total observed variance is heritability, or h^2. Estimation of heritability is often the major goal of efforts to partition phenotypic variance. If 40% of the observed variation in a trait (V_P) within a given population is tentatively attributed to additive variance (V_A), the heritability estimate of that trait is expressed as 0.40 (V_A/V_P). Some confidence limit, associated with the estimation procedure, is normally included.

Clearly, heritability is an integral concept in genetic improvement through selection. A big fish may be big due to any number of causes, many of which may exert more influence on its phenotype than the allelic average effects in its genotype. A low heritability, that is, relatively little additive genetic variance in relation to the total phenotypic variance, implies little genetic improvement can be made from one generation to the next through selection because too many other factors are contributing to the phenotypic variation we observe. Bigger may not always be better, just luckier in terms of environment or combination of parents. Once an animal or plant breeder derives an estimate of heritability, he or she can predict to some corresponding degree the performance of any individual's offspring based on that individual's phenotypic value.

Aquaculturists are typically interested in heritability estimates for various traits and species, but estimates of dominance variance produced through nested or factorial mating designs also provide valuable indications of the potential for improvement through crossbreeding. Impacts of dominance effects, as described previously, depend on specific interactions of alleles at individual loci, and these combinations are broken down when alleles separate to form gametes. The reinstitution of specific combinations cannot be predicted within a randomly breeding population. As a result, average effects and the heritability estimates ultimately derived from them are often the only thing practical aquaculture geneticists can hang their hats on when trying to make improvements from one generation to the next within a breeding population.

3.3 PRACTICE

3.3.1 Directed mating

As certain assumptions must be met in any analysis of variance, studies designed to partition phenotypic variation in aquatic species require careful control over culture conditions. Unfortunately, they additionally require substantial control over the timing and direction of matings between broodstock. This need often leads to logistical problems when dealing with certain aquatic species. If a number of family groups are to be compared over some period of early growth, an assumption must be made that age (and/or size, depending on the circumstances) does not differ significantly among groups at the start of the trial.

The biology of the species in question often determines what types of matings, and therefore analyses of variance, are possible. When gametes can be easily collected, stored and combined, as in many salmonid species, requirements for contemporaneous sib groups need not create serious problems (see Chapter 9). Zygotes can be formed relatively simultaneously, with any number of combinations possible. A given female can be "mated" to several males just as easily as the reverse situation. This makes a factorial design feasible, where each dam is mated to every sire and every sire is mated to each dam. It also makes nested designs, where two or more dams are mated to the same sire, much more precise by reducing or eliminating age differences among full-sib groups.

If matings must be allowed to take place naturally, for example in such species as prawns or certain catfish, the breeder may have to allow males to perform multiple matings at their own pace and settle for the most contemporaneous groupings they can obtain (Fig. 3.4). Alternately, it may be possible to analyze the data at regular intervals throughout the study to identify any significant effects due to age or size at the initiation of the study, some of which may then be statistically adjusted for. Even with artificially produced fish such as hybrid striped bass, matings and their synchronization will be limited by the rate at which the available female broodstock ovulate and become available for stripping.

Crustaceans in general pose difficulties when directed matings are required. In the case of freshwater crayfish, particular problems due to biology and natural history arise. Under normal conditions female crayfish may mate over time with several males, carrying the attached spermatophores until they are ready to ovulate in seclusion (Fig. 3.5). Lacking genetic markers such as coloration or albinism in both sexes, the contribution of each sire to a clutch often cannot be easily ascertained. One practical solution to allow directed mating of specific individuals is to scrub previously deposited spermatophores from the female in question with rubbing alcohol and allow her to mate once more with a designated sire. Unfortunately, even when this can be ac-

Fig. 3.4 Directed mating in *Macrobrachium acanthurus*. Note cast exoskeleton from female prenuptial molt near center of tank. Male has been marked with typewriter correction fluid for short-term identification. Screen serves to divide tank into two chambers, each with a mature female. (Photo by the author.)

complished, little progress has been made in identifying environmental cues (temperature and/or photoperiod manipulation) to trigger simultaneous ovulation and oviposition in many commercially cultured crayfish species. In these and many other species, if the resulting disparity in time of hatching among clutches is too extreme it can all but rule out a meaningful analysis even when directed matings have been successful.

3.3.2 Identifying or segregating family groups

As previously stated, offspring from specific matings must be identifiable in order to apply analyses that partition sources of variation. Unless offspring can be individually identified (say, with internal or external tags) and raised in a common environment, researchers must keep family groups in individual containment systems. Common environmental influences may artificially increase the resemblance between related individuals within a group and/or inflate differences between segregated family groups. To account for these differences in culture environments, each family group in turn should be divided into two or more tanks, net-pens, ponds, or whatever other containment unit is being used, to provide an estimate within the analysis for this source of environmental variance.

The assumption of identical environmental conditions for all offspring must include all the major variables such as stocking density, age and size at stocking, water quality, temperature, ration quality and

46 *Practical Genetics for Aquaculture*

Fig. 3.5 The reproductive biology of male (right) and female (left) cambarid crayfish prevents artificial spawning for all intents and purposes. Through the use of modified pleopods (seen at right extending forward from the base of the abdomen between the walking legs), the male deposits a spermatophore into the female's annulus ventralis (seen at left between the bases of the last 2 [fourth and fifth] pairs of walking legs), where it cures upon contact with water. As eggs are released weeks or months later through ducts at the bases of the third pair of walking legs (seen at left as small, circular plates), the spermatophore is dissolved to allow fertilization to occur as eggs pass downward to become attached to pleopods for incubation. (Photo courtesy J. Wozniak.)

quantity, exposure to disturbance, pathogens and predators, and so on (Fig. 3.6). Constancy of water quality may be the most difficult of these assumptions to envision. It is hard to say which type of culture environment typically in use would be the easiest or most difficult to replicate from one experimental system or generation to the next: flow-through, recirculating or static. Any of these types of systems possesses its own potential for variability.

When progeny from specific matings must be segregated to allow for identification, the chance of significant differences in water quality conditions among containment units increases (Fig. 3.7). Addition-

Fig. 3.6 Although replicated tanks within a system may appear to possess identical environmental conditions, subtle differences such as lighting or exposure to disturbance from foot traffic can create substantial bias in performance of family groups. (Photo by the author.)

ally, in many aquatic species some level of mortality over time is to be expected. When a population is subdivided, random mortality may result in substantial differences in stocking density among segregated containment units. If stocking density significantly affects the expression of the trait or traits in question, statistical adjustments may be required.

Another problem that can arise with segregation of family groups is competition. Conditions may arise which reduce or even overwhelm any resemblance between relatives within a specific group, due to competition for limited resources or establishment of social hierarchies (Fig. 3.8). If one individual gets more to eat, others will get less. In species with tendencies toward cannibalism at early life stages, such as many marine finfish, or in species that establish imposing behavioral

48   *Practical Genetics for Aquaculture*

Fig. 3.7 It is often possible to divide one or more tanks within a recirculating system into compartments for segregating family groups. If each group can be replicated in each of a number of tanks or systems, statistical analysis can be strengthened considerably. (Photo by the author.)

Fig. 3.8 Cages afford some degree of common environment while still allowing for segregation of family groups, but competition and social hierarchies can result in negative covariances. (Photo courtesy J.W. Avault, Jr.)

hierarchies, such as tilapia or freshwater prawns, additive variance for traits such as growth may be obscured when family groups are segregated. The extent to which culture conditions within a study should reflect those in the corresponding industry must be evaluated on a case by case basis, depending on the goals of the investigators and the intended application of their findings.

Marking techniques have become available to allow communal rearing of family groups of many species, eliminating some of the environmental bias pioneering researchers had to contend with (Fig. 3.9). In an attempt to quantify the extent to which these sources of bias can cause problems, Winkelman *et al.* (1992) used passive integrated transponder (PIT) tags to identify chinook salmon (*Oncorhynchus tshawytscha*) from full-sib families within three distinct strains. Fish from all families and strains were raised together in two separate sites, and growth in length and weight was recorded after 9 months of saltwater rearing. Two statistical models were used to analyze growth data: one which accounted for dominance effects and common environment and another which did not include these effects. Heritability estimates for the reduced model were notably inflated when compared with those produced with the full model.

Consider the typical life history of almost any aquacultured species. Random proximity to a food item in a larval culture tank or nursery pond could conceivably result in a life-long size advantage for individuals of certain species under certain culture conditions. To reiterate, at times bigger simply means luckier. This is why statistical analyses

Fig. 3.9 The tendency for periodic molting complicates marking in many crustaceans. In recent years, innovative techniques have been developed to improve survival associated with marking as well as marker retention. (Photo by the author.)

of large numbers of individual organisms are often required to obtain meaningful understanding of genetic variances in aquaculture stocks. Competition within sibling groups often renders individual offspring measurement useless in plant breeding analyses, and several researchers have voiced the opinion over the years that the way we culture many aquatic species is more comparable to plant production than traditional animal breeding. Perhaps more aquaculture genetic studies should reflect this point, utilizing measurements taken on groups of progeny rather than individual organisms. However, required increases in the size of culture units, in labor, and in technical support would render such an approach virtually impossible for all but the most advanced research facilities.

3.3.3 Constraints: analysis and interpretation

Behavioral interactions must be considered when determining grow-out conditions to be used for evaluating strains or family groups. Although it is always desirable to maintain identical environmental conditions for all groups being evaluated, when stocks or families are raised communally, competition or aggression may serve to mask actual performance potential. When a line of red tilapia (*Oreochromis*) were cultured in net-pens in green-water recirculating systems with each of four separate species or varieties (*O. aureus, O. niloticus, O. aureus*-based hybrids and genetically-male *O. niloticus*) they performed poorly in terms of growth and survival. When raised separately in the same systems in designated net-pens, however, the performance of these red tilapia was comparable to that of the fastest-growing line in the study (genetically-male *O. niloticus*) (Lutz, unpublished data).

A related constraint for many aquaculture geneticists is that most studies examining genetic control of metric characters require measurements from large numbers of individuals to attain acceptable levels of precision, even when problems arising from segregation of sibling groups and within-group competition can be avoided. Unfortunately, replicated culture facilities are generally in short supply at most hatcheries and research stations, limiting the number of broodstock that can be used in analyses of phenotypic variation. Additionally, if a metric character cannot be easily quantified or is physically difficult to measure accurately, progress is often slow and expensive.

Even with prepackaged statistical software, mathematical complications can bias traditional estimates of additive and dominance effects. Some sources of variation in ANOVA-based models cannot be considered random effects, although many popular software packages handle them as such. Increasingly, animal and plant breeders have accepted the fact that mixed-model analyses of variance, which produce simultaneous solutions for fixed and random effects, are more appropriate for these types of studies than traditional ANOVAs.

Fixed effects in an ANOVA model should include factors such as gender, particular hatchery or growout facilities (similar to the concept of a hatchery constant for salmonid production), distinct production seasons, or other similar effects for which Best Linear Unbiased Estimates (BLUEs) can be developed (Table 3.4). Modern statistical software incorporating mixed models can allow for the influences of sex and state of maturity to be estimated while still utilizing entire progeny data sets to estimate heritabilities. These approaches can help "clean up" or at least address some of the bias that often plagues mixed-sex heritability experiments (Table 3.5). An added benefit in the use of mixed model analysis is the calculation of Best Linear Unbiased Predictors (BLUPs) of broodstock (at least those broodstock that produce more than one set of siblings in an analysis) when mated randomly within the population. An excellent illustrative example of this analytical approach can be found in Rye and Refstie (1995).

Many research and production stocks are relatively isolated from generation to generation. Many more were initially established with relatively small breeding populations. If a Macrobrachium hatchery in Belgium reports progress in selecting for cold tolerance, a facility in California should not necessarily begin planning a breeding program. They may, however, want to plan an analysis to examine their particular population. One caveat in the application of variance analysis in aquaculture is the qualification that estimates of genetic and environmental variances, as well as heritability estimates, only apply to the populations from which the estimates were derived. They may not reflect genetic and environmental influences in some, or even most, aquaculture stocks and facilities and will almost certainly not reflect conditions in wild stocks.

One potential means to approximate a species-wide estimate of components of genetic variance would be to extend the factorial analysis approach to utilize inbred lines within a species rather than

Table 3.4 Best Linear Unbiased Estimates (BLUEs) and significance levels of the fixed effects of sex and male reproductive maturity on growth, body size, and processing traits in *Procambarus clarkii* 150 days post-stocking (after Lutz & Wolters 1999).

		Fixed effect levels		
			Males	
Trait	Mean	Females	Mature	Immature
Total length (mm)	88.558	0.263	1.506*	−1.769*
Total weight (g)	24.159	−1.358*	5.256**	−3.898**
Abdomen meat wt (g)	4.234	0.158**	−0.260**	0.102*
Dressout percentage	19.3	1.4**	−3.9**	2.5**
Carapace width (mm)	20.275	−0.208**	0.599**	−0.392**
Carapace length (mm)	47.374	−0.651**	0.652**	0.000
Abdomen depth (mm)	12.157	0.171**	−0.121**	−0.050
Abdomen width (mm)	17.531	0.391**	−0.068	−0.323**
Chela length (mm)	35.023	−4.552**	8.067**	−3.515**
Chela width (mm)	12.411	−0.902**	2.558**	−1.656**

*, ** : $P < 0.05, 0.01$, statistical significance.

Table 3.5 Comparison of heritability estimates, including standard errors, using ANOVA-based quadratics and Henderson's mixed model Method III quadratics for growth, body size, and processing traits in *Procambarus clarkii* 150 days post-stocking (after Lutz & Wolters 1999).

Trait	ANOVA			Method III		
	h^2_s	h^2_d	h^2_{s+d}	h^2_s	h^2_d	h^2_{s+d}
Total length	0.12 (0.22)	0.87** (0.26)	0.49** (0.13)	0.11 (0.08)	0.23* (0.11)	0.17* (0.07)
Total weight	0.01 (0.17)	0.77** (0.24)	0.39** (0.10)	0.05 (0.05)	0.21* (0.10)	0.13* (0.06)
Abdomen meat weight	0.21 (0.36)	1.32** (0.37)	0.77** (0.19)	0.23 (0.15)	1.49** (0.27)	0.86** (0.17)
Dressout %	0.38 (0.31)	0.71** (0.25)	0.54** (0.16)	0.62* (0.27)	0.72** (0.22)	0.67** (0.16)
Carapace width	0.10 (0.10)	0.17 (0.10)	0.13** (0.05)	0.04 (0.05)	0.20* (0.10)	0.12* (0.06)
Carapace length	−0.04 (0.09)	0.35* (0.14)	0.16** (0.05)	0.00 a	0.44** (0.14)	0.22* (0.08)
Abdomen depth	0.18 (0.13)	0.19 (0.10)	0.18* (0.09)	0.22 (0.11)	0.18 (0.10)	0.20** (0.07)
Abdomen width	0.04 (0.12)	0.41* (0.16)	0.22** (0.07)	0.02 (0.05)	0.46* (0.14)	0.24** (0.08)
Chela length	0.09 (0.09)	0.14 (0.09)	0.12* (0.05)	0.17 (0.10)	0.35** (0.13)	0.26** (0.09)
Chela width	0.17 (0.16)	0.41* (0.16)	0.29* (0.09)	0.24 (0.12)	0.65** (0.17)	0.45** (0.12)

a: Non-estimable
*,** $P < 0.05, 0.01$, statistical significance.

individual broodstock within a population. This type of approach, referred to as a diallel cross, provides the same valuable estimates of additive, dominance and maternal effects for the whole of the base population from which the inbred lines were derived. In some cases, when inbred lines reflect diverse geographic origins, this can be equated to the species as a whole, albeit only under the particular culture conditions employed.

3.4 NOTABLE INVESTIGATIONS AND APPLICATIONS

3.4.1 Interpreting and applying heritability estimates

Consider the development of a study to partition phenotypic variation associated with disease resistance. This is probably one of the most widespread areas of interest in aquaculture genetics, and the broad principles associated with this topic are still not fully explored. It might seem that if disease resistance were even moderately heritable, the survivors of any given outbreak would pass along superior resistance to their offspring, and, eventually, the disease would cease to be a consideration. In terms of evolutionary history, this may in fact have occurred in many instances, but genetic modification of the current balance of power between a particular pathogen and the immune system of its target species may or may not be feasible over a mere five to ten generations. If a nested or factorial mating design requires several or many siblings to be maintained in one tank and challenged with a pathogen, as time passes and the least-resistant individuals become infected, should the probability of infection go up or down for the more resistant individuals? How closely would this reflect a typical production environment, if at all? These are examples of the types of questions that must be reckoned with when setting up an experimental design to estimate genetic effects.

Yamamoto *et al.* (1991) from the Nagano Prefecture Fisheries Experiment Station in Nagano, Japan, reported on the heritability of resistance to infectious hematopoietic necrosis (IHNV) in rainbow trout (*Oncorhynchus mykiss*). Their experimental design and results provide an illustration of the importance of interpreting a heritability estimate in the context in which it was developed. The researchers set up a hierarchical (nested) mating design to produce full- and half-sib groups of juveniles, which were divided into two challenge studies. Juveniles were exposed to IHNV by dipping in low dose or high dose solutions, with resulting estimates of the heritability (h^2_s) of resistance of 0.05 and 0.51, respectively. If we remember that heritability is nothing more than a description of the portion of observed variation that can be attributed to additive variance, we can develop a straightforward interpretation of these estimates. Under low dose conditions, a number of factors other than additive variance influenced individual susceptibility. At the high dosage level, however, these factors apparently played a less important role compared to additive genetic attributes.

In the area of reproductive traits, aquaculturists are often concerned with the performance of the individual, such as age at first spawning, fertility, etc. Alternately, we may be concerned with the performance of an individual's offspring. Fecundity of an individual female may be negatively correlated with the early life history performance of its offspring due to individual egg and offspring size. Accordingly, the number of viable post-larvae or fry weaned onto an artificial diet might be a better indicator of an individual female's reproductive performance in many circumstances.

The early growth and survival of most aquaculture species represents a gradual transition from the maternal influences of egg size, egg quality and, occasionally, suitability of the hatching environment to the expression of individual genomes as development progresses. On the other hand, as time passes the cumulative impact of non-genetic environmental factors can overwhelm additive genetic influences (Fig. 3.10). Wada (1989) reported on his estimation of the heritability of larval shell length in the Japanese pearl oyster. Two factorial matings of five females by four males were used to produce larvae which were measured for shell length on days 4–5, 10 and 15. Average heritability

Fig. 3.10 Over time, density effects can confound genetic influences when family groups are segregated, even with identical water quality from flow-through or recirculating supplies. (Photo by the author.)

(h^2_s) estimates from the three sampling ages were 0.335, 0.181 and 0.078, respectively. In this instance the data suggest that as time passed, factors other than additive genetic effects played a greater and greater role in phenotypic variation in shell length.

Selection for market-related characters is predicated on attaining acceptable production performance utilizing available technology. The goal is to take a production process that already works and enhance the end-product to conform to consumer demand. As an example, the heritability of flesh coloration in pen-reared coho salmon was reported on by Iwamoto *et al.* (1990). When fish from two separate brood years were fed a diet supplemented with canthaxanthin, estimated heritabilities (h^2_s) for carotenoid levels in the flesh were 0.50 ± 0.16 and 0.30 ± 0.14. These levels, notably, were positively correlated with weight and therefore most likely with food consumption.

3.4.2 A case study: *Ictalurus punctatus*

Simply designing and setting up an experiment to allow for growth data to be put into an analysis of variance is a formidable undertaking when working around the reproductive and behavioral peculiarities of many aquaculture species. Reagan *et al.* (1976) reported on efforts to estimate the heritability of growth rate in channel catfish. The first step in their initiative was to complete a nested mating design utilizing 20 males, each mated to two females, resulting in 40 full-sib and 20 half-sib families of fry. Since male channel catfish cannot be strip spawned (due to a convoluted, multiple-lobed testis), the researchers were forced to encourage natural matings over as small a time span as possible. The aggressive behavior associated with mating in this species required a substantial deviation from random mating: pairing of similar-sized broodfish in spawning pens. This practice, known as assortative mating, was considered necessary to reduce courtship-related injuries and mortality.

Once the matings were completed, more real-world problems disrupted the experimental design. Three full-sib groups were lost, leaving the remaining full-sib groups from the three sires in question without half-sib counterparts. As a result, the overall analysis was reduced to include data from the offspring of only 17 sires and 34 dams. Another problem, requiring statistical adjustment, was variable stocking density. The researchers were forced to use cages to keep full-sib groups contained and identifiable, and stocking density varied greatly as a result of differential (and, for the sake of the analysis, random) mortality. Stocking density was reduced to 35 fish per cage (or less, in those groups that already had less than 35 fish) 5 months into the study. Statistical adjustments through covariance correction equations were then made to allow comparisons of family groups raised under different cage densities prior to (and after) the density reduction. In spite of these problems and possible bias resulting from common environmental effects, assortative mating and variation in feeding rates, these researchers developed valuable initial estimates of heritabilities in channel catfish (Table 3.6). Their work also shed light on pitfalls to be avoided in future variance partitioning studies with this and other species.

Table 3.6 Heritability (h^2_s) estimates for growth: practical trials with warmwater species.

Species	Trait	Estimate	Source*
I. punctatus	Length (5 mo.)	0.12 ± 0.30	Reagan *et al.*
	Length (15 mo.)	0.67 ± 0.57	Reagan *et al.*
	Length (48 mo.)		
	males	0.39 ± 0.39	El-Ibiary and Joyce
	females	0.81 ± 0.46	El-Ibiary and Joyce
	Weight (5 mo.)	0.61 ± 0.35	Reagan *et al.*
	Weight (15 mo.)	0.75 ± 0.53	Reagan *et al.*
	Weight (48 mo.)		
	males	0.27 ± 0.37	El-Ibiary and Joyce
	females	0.52 ± 0.42	El-Ibiary and Joyce
M. rosenbergii	Weight (70 days)	0.10 ± 0.07	Malecha *et al.*
	Weight (119 days)	0.19 + 0.11	Malecha *et al.*
	Weight (238 days)	0.14 ± 0.11	Malecha *et al.*
	Weight (238 days)	0.18 ± 0.12	Malecha *et al.*
	Weight (311 days)	0.16 ± 0.14	Malecha *et al.*
P. clarkii	Length (150 days)	0.18 ± 0.25	Lutz and Wolters
	Weight (150 days)	0.08 ± 0.20	Lutz and Wolters

* Citations listed in text pages 54–57

Not long after Reagan *et al.* (1976) reported their results, El-Ibiary and Joyce (1978) reported on similar, but unrelated, efforts to estimate heritabilities in channel catfish. They examined nine females and nine males from each of 26 full-sib families nested within 13 half-sib families. Sire- and dam-based heritabilities were calculated for males and females separately for a number of body size and slaughter traits at an age of 48 weeks post-hatching. In most traits measured, dam-based heritabilities were higher, suggesting influences from dominance effects or common environment. Among sire-based heritabilities, values were moderate to high for body weight and total length (Table 3.6), and generally very low for dressing percentage, head weight percentage and percent lipid content. Large standard errors were associated with all estimates due to limited degrees of freedom.

3.4.3 A case study: *Macrobrachium rosenbergii*

Malecha *et al.* (1984) described procedures utilized to estimate the heritability of growth patterns in juvenile *Macrobrachium rosenbergii*. Again, constraints related to culture methods and reproductive and behavioral characteristics of the species in question transformed what began as a fairly straightforward experiment into a complex attempt to deal with various sources of experimental bias. The authors' perseverance and recommendations serve as examples to others contemplating similar undertakings.

Over a 7-day period, the authors were able to arrange for the production of 50 full-sib families from 16 separate sires. Female prawns cannot be injected with hormones to hasten ovulation and mating, and milt cannot be stripped from males. Quite the contrary: prawns must be relied upon to act out a behavioral/mechanical sequence in which the male expresses a gelatinous spermatophore and places it ventrally on the female's cephalothorax shortly after she completes her pre-spawning molt. Individual male prawns, being fairly agreeable to this sort of thing, were mated successfully in this way with two to five dams each. The close timing of these matings was possible by monitoring the pre-ecdysis nuptial molt stage of the females (dams) to be used in the study.

Producing usable full-sib families, however, was no small feat. Only 154 of 316 dams originally collected were usable. Nineteen of these failed to molt, 18 post-molt matings failed entirely, 3 of 57 spawns designated for inclusion in the study were lost, and three families suffered very high levels of mortality for unexplained reasons subsequent to hatching. Great pains were taken to eliminate sources of environmental bias from the experimental design. Each full-sib family of larvae was divided into two groups upon hatching, and all larval groups were reared in screen bags in two large green-water tanks. After metamorphosis, samples of juveniles from each family were raised in individual compartments on water tables with a common water supply, and weighed at 70, 199, 238 and 311 days after being transferred.

Since water tables became a significant statistical effect in later measurements, heritabilities were estimated separately for each table because the unbalanced design (unequal numbers of dams per sire) posed complications in separating out table effects. Family-by-table interactions were not significant, so this approach provided a somewhat broader idea of the range of heritability that might be encountered in the population in question, but between the third and fourth measurement dates one water table was completely lost as the result of a technician's error.

Heritability estimates that were obtained, in spite of the technical problems, appeared to be sexually dimorphic as early as the 70th day after stocking. Estimates were developed for both sexes separately and combined. The authors speculated that factors relating to negative covariance within a common environment and the bimodal growth patterns typical of male *M. rosenbergii* may have somehow been associated with the much lower estimates derived for males. The authors concluded that male size in *M. rosenbergii* "may be a sex-limited secondary sexual characteristic under natural sexual selection" and that "such characteristics would be expected to have high genetic fitness and little genetic variance." Assortative mating may have been an additional source of bias in these heritability estimates also, since, as in the Reagan *et al.* (1976) study, dams and sires were matched to some degree based on size.

3.4.4 A case study: *Procambarus clarkii*

Lutz and Wolters (1989) reported on a similar attempt to estimate heritability of growth rate and a number of body size and processing traits in the red swamp crawfish, *Procambarus clarkii*. Red swamp crawfish proved to be much more difficult to breed in the laboratory than *Macrobrachium*. Cambarid crayfish do not exhibit a corresponding pre-nuptial molt. In fact, once amplexus has occurred, a female may carry one or more spermatophores for months before she decides the time is right for oviposition.

Utilizing a pool of 200 females and 38 males, efforts were made to mate two or more females to each male (Fig. 3.11). Successful oviposition resulted from only 63 of 190 matings, and egg mortality prevented the use of many clutches in the analysis (Fig. 3.12). Additionally, all but one of the first 14 spawns were consumed by their mothers during the first weeks of the study. Only 41 matings represented two or more females mated to a common male, and one of these clutches of eggs was produced much earlier than all the others, and so was discarded. Another clutch did not contain enough surviving individuals for stocking. The authors were, however, able to attain 39 full-sib families of similar age and size, produced by 14 sires each mated to two or more dams.

Growout units were intended to closely mimic commercial culture conditions. Families were raised in fiberglass pools with soil and rice stubble to serve as the basis of a detrital food chain. Once approximately

Fig. 3.11 Directed matings in crustacean species often rely on encouraging natural copulation and oviposition. Seen here are a male (above) and female red swamp crawfish (*Procambarus clarkii*). Note coin (US dime) for perspective. (Photo by the author.)

Fig. 3.12 Many freshwater and marine crustaceans fertilize their eggs as they are laid, utilizing previously-acquired spermatophores. Individual variation in maternal maintenance and care for eggs prior to hatching, which may have some genetic basis, can often result in additional phenotypic variance and confound estimates of additive or dominance genetic effects. (Photo courtesy J.W. Avault, Jr.)

half the males had molted to a reproductively mature state, growout was terminated. As in many other studies with this species, survival within families varied widely (3%–100%) and apparently randomly, since no significant family effects were apparent for survival. A total of 747 animals was harvested, but only 635 were available for analysis as a result of handling injuries and subsequent cannibalism. Like freshwater prawns, crawfish males exhibit substantial variation in their growth rates and body conformation as a result of differential timing of sexual maturity. Accordingly, heritabilities had to be estimated for all animals combined, for females separately, and for mature and immature males separately. For all animals combined, heritabilities were generally low to moderate, and dominance effects, confounding maternal effects and common environmental effects probably all served to bias estimates (Tables 3.5 and 3.6).

3.4.5 A case study: *Sparus aurata*

In certain aquatic species, simple life history and reproductive strategies can result in almost insurmountable constraints to directed matings. Gorshkov *et al.* (1997) reviewed the technical difficulties in producing genetically related groups (full- and/or half-sib families) in the gilthead seabream *Sparus aurata*. Inasmuch as this species is normally a group spawner, with females undergoing asynchronous daily development of oocytes over a 3-month period, specific matings of male and female broodstock have proven difficult to arrange. Another distinctive characteristic of gilthead is their protandrous hermaphroditic nature, undergoing a change in sex from male to female as they age.

In order to produce specific family groups, the authors utilized slow-release implants of GnRHa to promote final oocyte maturation and spawning in female broodstock. Individual females were also implanted with passive integrated transponder tags to allow for identification. Three approaches to production of families with known parentage were attempted: strip spawning, progeny testing, and single pair mating. In the strip-spawning trials, groups of fish consisting of five females and three males were held in maturation tanks and subsequently strip spawned to produce combinations of eggs (when available) and sperm from individual fish, resulting in a number of full- and half-sib family groups. In the progeny testing study, individual males were each housed with 15 females in separate spawning tanks. Spawning was allowed to take place naturally, and fertilized eggs representing half-sib groups were collected from each spawning tank. In the single pair mating trials, pairs of broodstock were held in partitioned tanks. Eggs, when produced, represented full-sib groups. Broodstock were rotated when spawning did not take place, resulting in new partners for previously unproductive fish. Attempts to induce single-pair matings involved a pool of 28 females and 22 males.

Of 52 attempts to produce full- and half-sib family groups through strip-spawning, only five were successful. Lack of ovulated eggs and poor egg quality when ovulation did take place were major drawbacks to the use of hand stripping in gilthead seabream. The authors noted that each female appeared to exhibit its own particular time of spawning on a daily basis, and the procedure of strip-spawning all females at the same time on any given day may have resulted in loss of potential spawns that would have been possible to obtain earlier or later in the day, depending on the preferred spawning time for each female. They concluded that one possible approach to improve the efficiency of strip spawning would be to adapt stripping times for individual females, although this would increase labor requirements substantially.

Ten of 23 attempts to conduct progeny testing, crossing one male with 15 females, proved successful. Again, failures were attributed to low spawning, fertilization and hatching rates. While this rate of success compares favorably to directed mating approaches used for many other aquatic species, the authors pointed out that relative contributions of females spawned with a given male could not be quantified. Another problem involved temporal disparity in production of half-sib groups, resulting in offspring of varying ages. Thirteen of 58 attempts to produce single-pair spawns succeeded. Many females failed to produce eggs of any quality, and rotation of broodstock to enhance compatibility may inadvertently have resulted in additional stress and associated suppression of oocyte maturation.

In spite of the low numbers of family groups available for analysis, the authors were able to document substantial genetic influences on weight gain in this species. Sires accounted for approximately 14% of total observed variance in weight gain by the end of growout trials for half-sib groups produced through progeny testing. Similarly, when growth of fingerlings produced through single-pair matings was evaluated, sires accounted for 29% of the variance in final weights. Unfortunately, the findings presented suggested that due to its high preference for group spawning, production of high numbers of family groups in *Sparus aurata* is probably impractical. As a result, improvement of production traits in this species might best be pursued through mass selection.

3.4.6 Growth, survival, conformation and dressout traits

When reviewing past studies of heritability in aquaculture species, it becomes apparent that growth rate is probably the most investigated production trait. Why do we usually practice selection for growth rate and not feed conversion? For a given fish or shrimp or other production animal, each day in the system requires maintenance of body weight. Over time, this can be a tremendous cumulative feed use. Complicating factors arise that have to be considered when assessing heritability of growth rate: if only so much feed can go into a system on a per day basis and faster growth requires more feed consumption, stocking rates of faster growing fish must be reduced while turnover within the system will increase.

Salmonid examples

All sorts of traits relating to growth and life history have been examined through variance partitioning studies. Precocious maturation can be a problem in the culture of some salmon, so Silverstein and Hershberger (1992) utilized a nested mating design to examine the heritability of this tendency. Using 40 full-sib coho salmon (*Oncorhynchus kisutch*) families representing paternal half-sib groups, they estimated a sire-based heritability of 0.05 + 0.05 and a dam-based heritability of 0.13 ± 0.11. Clearly, little influence from additive genetic variance was apparent, and little more could be attributed to dominance or maternal effects. A significant correlation, however, was found between the proportion of a family maturing precociously and egg size (0.40, $P = 0.01$). Altogether, these results discouraged attempting to select against precocious maturation in coho salmon.

When applying analysis of variance to estimate genetic and environmental effects, precision can be greatly improved if large numbers of full- and half-sib groups are available. Jonasson (1993) presented heritability estimates for Atlantic salmon (*Salmo salar*) based on numbers that would make many researchers envious. Survival of fry from a total of 298 full-sib families, representing roughly 100 half-sib groups, was recorded weekly for 12 weeks after first feeding. Two year classes from five distinct stocks were examined. A sample of fish from each family was subsequently measured for length and weight 190 days after first feeding. Significant stock effects were noted for survival, length and weight in both year classes. Individual weights and lengths were statistically adjusted for egg sizes and numbers of fish per family tanks. Sire heritabilities for adjusted length and weight were reported as 0.10 and 0.16, respectively, and the corresponding dam-based estimates were 0.39 and 0.36, suggesting significant dominance and/or maternal effects (other than egg size). Sire and dam heritability estimates for survival were calculated as 0.04 and 0.34, respectively, again implicating dominance, maternal or other non-additive influences.

When reproductive biology allows, factorial mating designs can provide more information per full-sib family produced than any other variance analysis approach. Nilsson (1994) reported results from a factorial mating trial using Arctic charr (*Salvelinus alpinus*). Survival to 7 months of age exhibited a heritability close to 0.0, but heritabilities for length and weight at 12 and 18 months of age ranged between 0.13 and 0.17. When the relationship between length and weight was presented as condition factor, heritabilities for this trait were 0.54 at 12 months and 0.32 at 18 months. However, genetic correlations between condition factor and length or weight measurements were comparatively low.

Occasionally, product traits such as flesh coloration become important in selection programs. Much interest has been directed to the degree of flesh color in salmon, and to its relationship to growth. Utilizing a mating scheme that included five males and ten females from each of six distinct coho salmon (*Oncorhynchus kisutch*) populations, Withler and Beacham (1994) estimated a genetic correlation of 0.4 ± 0.5 between flesh coloration and final weight. Heritability estimates were 0.6 ± 0.3 for final weight and 0.2 ± 0.1 for flesh coloration. The authors cited substantial variation in growth, survival and flesh coloration and emphasized the importance of utilizing superior founding populations in domestication and selection programs for this species.

One benefit of a nested mating design is the opportunity for comparisons of sire- and dam-based estimates of heritability, which in theory should provide some indication of the relative influence of dominance effects, maternal effects and common environment on the trait(s) in question. Occasionally, however, these distinct estimates may indicate little or no influence from non-additive sources. Rye and Refstie (1995) used a nested mating design to examine genetic and environmental parameters affecting growth and body size traits in Atlantic salmon (*Salmo salar*). Records were obtained on 1271 salmon representing 58 sires and 171 dams. Traits that described overall body size, including gutted and ungutted weight, length, circumference and condition factor all exhibited sire heritabilities in the range of 0.30–0.38. Similar values were determined for dam-based heritabilities, suggesting little impact from dominance or environmental influences. Statistical models indicated substantial effects of gender and state of maturity on growth traits, and the authors stressed the need for statistical accommodation of these effects when developing estimates of genetic variance and covariance.

As alluded to previously, genetic influences on a trait such as growth may be most notable early in life history, only to be obscured over time by random environmental effects. Or, on the other hand, as time passes and an organism grows and matures, its genetic merit may tend more to come to the forefront, allowing for a better estimating of breeding value over time. In growth studies with coho salmon (*Oncorhynchus kisutch*), Silverstein and Hershberger (1994) indicated that heritability estimates for growth-related characters were moderate to high (0.5 to 0.7) during the pre-smoltification period and only low to moderate (0.2 to 0.5) in post-smoltification sampling. This tendency was increased in a population that had been moved to sea water. In turn, genetic correlations between pre- and post-smoltification size were higher for the population reared in fresh water. Apparently, the sea-water environment resulted in increased environmental variation in the phenotype, reducing the portions of both additive genetic variance and covariance.

Crayfish examples

Working with the red claw crayfish *Cherax quadricarinatus*, Gu and coworkers (1995) determined that although significant growth differences among three stocks increased with age, family variation in growth within all stocks decreased over time. Heritability estimates based on full-sib groups were 0.72 ± 0.14, 0.48 ± 0.10, and 0.32 ± 0.31 at 30, 60 and 90 days post release, respectively. Clearly, factors other than additive genetic effects played an increasing role in phenotypic size as time progressed.

Bosworth et al. (1994) undertook a diallel cross among three populations of the red swamp crawfish (*Procambarus clarkii*) to estimate heterosis, line effects, maternal effects, and general combining ability for body size traits and dressout percentage. Estimates were made for all animals combined and separately for females, immature males and mature males (which differ significantly morphologically upon molting to the mature state). Data analysis suggested significant heterosis and line effects for dressout percentage, and significant maternal effects for body size traits in females and mature males suggested differences in egg quality or possibly cytoplasmic inheritance.

3.4.7 Disease resistance

Until recent years, furunculosis was one of the more serious diseases faced by salmon farmers in Europe. Gjedrem et al. (1991) reported on estimation of heritability in Atlantic salmon (*Salmo salar*) for susceptibility to furunculosis. They infected 5000 fish from 50 full-sib families within 25 paternal half-sib groups with *Aeromonas salmonicida*. The overall death rate averaged 68%, but family groups exhibited large variations in survival. Sire- and dam-based heritability estimates were 0.48 ± 0.17 and 0.32 ± 0.10, respectively, leading the authors to conclude that selection would be a practical means to improve resistance to furunculosis in this species.

Clearly, physiological stress and stress responses are implicated in the susceptibility of many aquatic species to disease. In a very large study of cortisol and glucose stress response in salmonids, Fevolden et al. (1993) examined 553 full-sib Atlantic salmon (*Salmo salar*) families and 281 full-sib rainbow trout (*Oncorhynchus mykiss*) families, nested within half-sib groups. Following exposure to a standardized confinement stress, plasma cortisol and glucose levels were sampled, revealing much higher levels of cortisol in Atlantic salmon than in rainbow trout. Glucose levels were more similar between species. Mean heritability estimates across year classes for cortisol and glucose levels were 0.05 and 0.03, respectively in *S. salar* and 0.27 and 0.07, respectively, in *O. mykiss*.

Lund et al. (1995) examined a number of immune parameters in 77 full-sib groups of Atlantic salmon (*Salmo salar*). The goal was to identify measurable components for use in indirect selection to improve disease resistance. Of the parameters studied, however, only lysozyme activity appeared to have substantial genetic variation and to be genetically associated with survival rates.

An examination of genetic and environmental effects influencing antibody response to Vibrio antigens in Atlantic salmon (*Salmo salar*) was presented by Fjalestad et al. (1996). They produced and immunized 57 full-sib families within 20 paternal half-sib families. Subsequently they sampled serum three different times from 1200 fish to estimate heritabilities for antibody concentrations. Estimates ranged from 0.00 to 0.10, with standard errors as high as 0.21. Variation in survival of progeny from different sires was significant, and a positive correlation was determined between mean survival of half-sib groups and antibody concentrations prior to immunization.

Wolters and Johnson (1995) presented results from a diallel cross of three strains of channel catfish (*Ictalurus punctatus*) to estimate genetic and environmental effects associated with resistance to the bacterium *Edwardsiella ictaluri*. Juveniles from each of nine possible crosses were evaluated for survival over a 28-day period following an immersion challenge. Survival ranged from 35.8% to 90.0%, indicating substantial genetic variation associated with resistance. Estimates of average, line and specific heterosis were all non-significant, but significant line effects were exhibited by two pure strains, suggesting additive genetic control over resistance. Significant maternal effects were also determined for one strain.

3.5 REFERENCES

Becker, W.A. (1984) *Manual of Quantitative Genetics*. Academic Enterprises, Pullman, Washington.

Bosworth, B., Wolters, W.R. & Saxton, A.M. (1994) Analysis of a diallel cross to estimate effects of crossing on performance of red swamp crawfish, *Procambarus clarkii*. *Aquaculture*, **121** (Suppl. 4), 301–312.

Danzmann, R.G. & Ferguson, M.M. (1995) Heterogeneity in the body size of Ontario cultured rainbow trout with different mitochondrial DNA haplotypes. *Aquaculture*, **137** (Suppl. 1–4), 231–244.

El-Ibiary, H.M. & Joyce, J.A. (1978) Heritability of body size traits, dressing weight and lipid content in channel catfish. *Journal of Animal Science*, **47** (Suppl. 1), 82–88.

Fevolden, S.E., Refstie, T. & Gjerde, B. (1993) Genetic and phenotypic parameters for cortisol and glucose stress response in Atlantic salmon and rainbow trout. *Aquaculture*, **118** (Suppl. 3–4), 205–216.

Fjalestad, K.T., Larsen, H.J.S. & Roeed, K.H. (1996) Antibody response in Atlantic salmon (*Salmo salar*) against *Vibrio anguillarum* and *Vibrio salmonicida* O-antigens: Heritabilities, genetic correlations and correlations with survival. *Aquaculture*, **145** (Suppl. 1–4), 77–89.

Gjedrem, T., Salte, R. & Gjoen, H. (1991) Genetic variation in susceptibility of Atlantic salmon to furunculosis. *Aquaculture*, **97** (Suppl. 1), 1–6.

Gorshkov, S., Gordin, H., Gorshkova, G. & Knibb, W. (1997) Reproductive constraints for family selection of the gilthead seabream (*Sparus aurata* L.). *The Israeli Journal of Aquaculture – Bamidgeh*, **49** (Suppl. 3), 124–134.

Gu, H., Mather, P.B. & Capra, M.F. (1995) Juvenile growth performance among stocks and families of red claw crayfish, *Cherax quadricarinatus* (von Martens). *Aquaculture*, **134** (Suppl. 1–2), 29–36.

Iwamoto, R.N., Meyers, J.M. & Hershberger, W.K. (1990) Heritability and genetic correlations for flesh coloration in pen-reared coho salmon. *Aquaculture*, **86** (Suppl. 2–3), 181–190.

Jonasson, J. (1993) Selection experiments in salmon ranching. I. Genetic and environmental sources of variation in survival and growth in freshwater. *Aquaculture*, **109** (Suppl. 3–4), 225–236.

Lund, T., Gjedrem, T., Bentsen, H.B., Eide, D.M., Larsen, H.J.S. & Roed, K.H. (1995) Genetic variation in immune parameters and associations to survival in Atlantic salmon. *Journal of Fisheries Biology*, **46** (Suppl. 5), 748–758.

Lutz, C.G. & Wolters, W.R. (1989) Estimation of heritabilities for growth, body size and processing traits in red swamp crawfish, *Procambarus clarkii* (Girard). *Aquaculture*, **78**, 21–33.

Lutz, C.G. & Wolters, W.R. (1999) Mixed model estimation of genetic and environmental correlations for body size and processing traits in red swamp crawfish, *Procambarus clarkii* (Girard). *Aquaculture Research*, **30**, 153–163.

Malecha, S.R., Masuno, S. & Onizuka, D. (1984) The feasibility of measuring the heritability of growth pattern variation in juvenile freshwater prawns, *Macrobrachium rosenbergii* (De Man). *Aquaculture*, **38**, 347–363.

Nilsson, J. (1994) Genetics of growth of juvenile Arctic char. *Transactions of the American Fisheries Society*, **123** (Suppl. 3), 430–434.

Reagan, R.E., Pardue, G.B. & Eisen, E.J. (1976) Predicting selection response for growth of channel catfish. *The Journal of Heredity*, **67**, 49–53.

Rye, M. & Refstie, T. (1995) Phenotypic and genetic parameters of body size traits in Atlantic salmon *Salmo salar* L. *Aquaculture Research*, **26** (Suppl. 12), 875–885.

Silverstein, J.T. & Hershberger, W.K. (1992) Precocious maturation in coho salmon (*Oncorhynchus kisutch*): Estimation of heritability. *Bulletin of the Aquaculture Association of Canada*, **92** (Suppl. 3), 34–36.

Silverstein, J.T. & Hershberger, W.K. (1994) Genetic parameters of size pre- and post-smoltification in coho salmon (*Oncorhynchus kisutch*). *Aquaculture*, **128** (Suppl. 1–2), 67–77.

Tave, D. (1991) Genetics of body color in tilapia. *Aquaculture Magazine*, **17** (Suppl. 2), 76–79.

Wada, K.T. (1989) Heritability estimation of larval shell length of the Japanese pearl oyster. *Bulletin of the National Research Institute of Aquaculture (Japan)/Watarai-gun*, **16**, 83–87.

Winkelman, A.M., Peterson, R.G. & Harrower, W.L. (1992) Dominance genetic effects and tank effects for weight and length in chinook salmon. *Bulletin of the Aquaculture Association of Canada*, **92** (Suppl. 3), 40–42.

Withler, R.E. & Beacham, T.D. (1994) Genetic variation in body weight and flesh colour of the coho salmon (*Oncorhynchus kisutch*) in British Columbia. *Aquaculture*, **119** (Suppl. 2–3), 135–148.

Wolters, W.R. & Johnson, M.R. (1995) Analysis of a diallel cross to estimate effects of crossing on resistance to enteric septicemia in channel catfish, *Ictalurus punctatus*. *Aquaculture*, **137** (Suppl. 1–4), 263–269.

Yamamoto, S., Sanjyo, I., Sato, R., Kohara, M. & Tahara, H. (1991) Estimation of the heritability for disease resistance to infectious hematopoietic necrosis in rainbow trout. *Nippon Suisan Gakkaishi*, **57** (Suppl. 8), 1519–1522.

Chapter 4
Selection and Realized Heritability

4.1 INTRODUCTION

Virtually all of the progress made in genetic improvement of livestock until just a few years ago can be attributed to the application of selection pressure. The practice of selection relies on utilizing the most desirable individuals within a population to produce the subsequent generation, or at the very least on the exclusion of undesirable individuals from the breeding population. Of course, what is desirable will depend on what traits the breeder wishes to modify. Furthermore, how much progress can be made from generation to generation depends largely upon the extent to which the observed phenotypic variance reflects underlying genetic variance – particularly additive genetic variance.

When additive effects play a significant role in production traits, selection becomes a practical means of genetic improvement in almost any aquaculture operation. Some of the things aquaculturists may wish to improve through selection are obvious: disease resistance, reproductive performance, market-related characters and growth, but setting up an experiment to estimate actual or predicted selection response in such traits is often quite difficult. When individuals are retained for breeding purposes based solely on their individual performance, the practice is referred to as individual, or mass, selection. Other types of selection, based wholly or in part on the performance of entire families, are often practiced as well.

As discussed in the previous chapter, the degree to which genetic variance, or specifically additive genetic variance, influences the phenotype of a metric character is referred to as the heritability of the trait, usually expressed as h^2. Several forms of heritability estimates are commonly used, depending on the types of data available for their calculation, and all of them take the form of a ratio between phenotypic variance attributable to genetic effects and total phenotypic variance, expressed as a value between 0.0 and 1.0. Understanding the heritability of a particular trait allows for the formulation of efficient breeding programs to improve performance in production stocks. Traits with moderate or low heritabilities often require somewhat different approaches than highly heritable ones.

4.2 THEORY

Perhaps the simplest form of selection relies on comparing various strains within a species and utilizing the best one(s) for culture purposes. In large part, this practice relies on identifying differences resulting from additive genetic effects, but no immediate effort is made to alter the genetic makeup of the strains in question. In most circumstances, however, we recognize selection as the process of altering, generation by generation, the genetic makeup of a population in order to produce more desirable phenotypes. In this sense, a somewhat inferior strain that possesses significant additive variance may be a better starting point for selection and domestication than a moderately superior population with little additive variance to exploit.

4.2.1 Estimating and predicting heritability

In contrast to single-generation variance partitioning procedures reviewed in the previous chapter, one of the simplest approaches to estimating heritability and predicted selection response relies on a form of analysis where offspring performance is evaluated statistically in relation to parental performance. Clearly, the ability to produce, culture and track the performance of large numbers of full- and half-sib family groups is beyond the means of most commercial fish farms, and many research facilities as well. If progeny can only be identified as a pooled group of offspring from a pooled group of broodstock, the analysis of selection response is usually not too powerful, but if individual offspring can be identified as the progeny of specific broodstock, the resulting estimates can become much more precise, and individual breeding values can even be estimated.

For virtually any metric trait, especially in aquaculture where culture environments can be highly variable in many phases of the life cycle, only a portion of an individual's phenotypic superiority or inferiority will be due to additive genetic effects, or any genetic effects for that matter (Fig. 4.1). One result of this truth is that even though two select individuals may be superior to the population mean from their own generation, the overall performance of their offspring usually falls somewhere between that population mean and the performance of the selected parents. The analysis of this tendency for performance to "regress" toward the population mean with successive generations was the basis for the statistical term regression. In many selection trials at commercial and state-run facilities aquaculturists have mistakenly compared offspring performance with that of the offspring's selected parents, rather than with the overall population mean for the previous generation, and erroneously concluded that selection is not worth pursuing.

As outlined in the previous chapter, there are several components of genetic variation other than additive effects. These include dominance effects and several types of interactions among the loci involved in the

Fig. 4.1 Phenotypic variation, as seen in these red drum (*Sciaenops ocellatus*), may be more the result of environmental variation than of additive genetic variance which serves as the basis for improvement through selection. (Photo by the author.)

expression of any particular trait. At times, reproductive biology, mating designs or available facilities will not allow these components to be entirely isolated. As a result, various types of heritability estimates can be found in the literature. Heritability "in the broad sense" refers to an estimate of the total genotypic variance in relation to the phenotypic variance. Broad sense heritability can be represented as:

$$h^2_B = (V_A + V_D + V_I) / (V_A + V_D + V_I + V_M + V_E).$$

In contrast, heritability in the narrow sense focuses solely on the contribution of additive genetic variance to the phenotypic variance we observe. It is generally expressed as:

$$h^2_N = V_A / (V_A + V_D + V_I + V_M + V_E),$$

or as

$$V_A / (V_A + V_E') = V_A / V_P,$$

where V_E' represents all phenotypic variance not directly attributable to additive genetic effects. It is generally assumed that each of these components of variance is independent. This might not be the case, however, if selected stocks were somehow provided with a superior environment. In such a situation, the covariance of A and E' would not equal 0.

Fortunately, most available estimates of heritability are somewhat more precise than h^2_B, but none are as unbiased as h^2_N. Clearly, the value of h^2_N lies in its ability to predict that portion of the phenotypic differences in breeding (or potentially breeding) individuals that will be reflected in their offspring. In the past, many researchers have reported heritability estimates for aquatic species that include components other than just V(A) in the numerator, both genetic and environmental, without focusing on the actual composition of the estimates. This practice reduces the utility of these estimates to some degree, especially if they are not represented as upwardly biased.

As outlined above, heritability can be described and estimated as the regression of an individual's genotypic value for a given trait on its observed phenotypic value. In this way, a change in genotypic value relating to a known change in phenotypic value can be developed as follows:

$$h^2 = bA,P = \text{cov } A,P / V_P = \text{cov } A(A + E') / VP = V_A/V_P,$$

since the covariance of A and E' is again assumed to be 0.

It also follows that in real-world selection trials the degree to which selected individuals' offspring are superior to the population average of the entire parental generation should reflect how much of their parents' superiority was due to additive genetic effects. This of course requires that the two generations can be raised under sufficiently similar, if not identical, conditions. Data such as food conversion efficiency, fecundity, growth rate, egg fertility, survival, dressout percentage, lethal low temperature, etc. can be evaluated as paired observations. A pair of observations can include a single parental value or a mid-parent (average) value, in conjunction with a single offspring value or a mean derived from many progeny. The degree of resemblance between these types of relatives is reflected in the regression of offspring values on parental values, and their covariance is calculated from the sum of cross-products.

Through mathematical deductions too tedious for this discussion, quantitative geneticists long ago determined that the genetic covariance between offspring and parental values represents half of the additive genetic variance within the population. If the breeding value, or genotype of an individual is represented as A, the individual will pass along A/2 to its offspring. The offsprings' phenotypes, in turn, can be represented as A + E'. Recalling that A and E' are independent of each other, the covariance between the individual and its offspring can be derived as:

$$\text{cov } [A/2][A + E'] = V_A/2 + 0/2 = V_A/2, \text{ since cov } AE' = 0.$$

As a result, when offspring values are regressed on those of a single parent, the resulting regression coefficient (b) is equal to ½ h^2 [$V_A/2$ / V_P]. If the regression utilizes mid-parent values derived from both parents, the denominator of the calculation, V_p, is also halved and the b coefficient equals h^2.

If phenotypic variance differs substantially between sexes, heritability regressions must be calculated for each sex separately. Male parent values are paired directly with those of their male offspring and, additionally, with those of their female offspring after each female offspring value is multiplied by the ratio of the phenotypic standard deviation of male progeny over the phenotypic standard deviation of female progeny. The same procedure applies to the regression of offspring on female parent values, with male offspring measurements being multiplied by the ratio of the female progeny phenotypic standard deviation to the male progeny phenotypic standard deviation. Sex-linked inheritance is usually assumed to be of minor importance in many studies of heritability, and so we will not address it in any detail in this discussion, but it can be a consideration in organisms with relatively few chromosomes or where sex-determining chromosomes are comparatively large.

Most heritability estimates will have some variance associated with them, but extremely large numbers of observations are required to obtain much precision. Errors of estimates derived from parent–offspring regressions can be derived from the error of a regression coefficient as follows:

$$V(b) = [1/(N-2)][\{V(Y)/V(X)\} - b2] \text{ AND } V(h^2) = 4V(b)$$

where Y represents offspring performance, X represents parent performance, N is the number of paired observations and b is the regression coefficient.

4.2.2 Applying selection

The grave limitation of any heritability estimate is its applicability, or lack thereof. In theory, h^2 is merely a descriptive attribute of the population from which it was estimated. Additionally, it reflects that particular population raised under the specific environmental conditions that prevailed during the period of time in which measurements were recorded. In many ways, a heritability estimate is like a photograph – the subject population may or may not be easily recognized the next time it is examined, depending on any number of factors.

Now that some groundwork is in place regarding the concept of heritability, it applications and manifestations bear discussion. The practice of selection is based on heritability, both conceptually and mathematically. Consider this example. A population of catfish exhibits an overall mean weight at 2 years of age of 2.10 kg. A breeder selects the best-performing individuals to serve as broodstock to produce the next season's fingerlings. The mean weight of the selected broodstock is 2.45 kg. The difference between these two values, 2.45 kg and 2.10 kg, is termed the selection differential, and the expected breeding value of the selected individuals will be (h^2)(0.35) kg above the population mean of 2.10. This is also the expected mean of the selected individuals' progeny, under identical environmental conditions.

When a population is arbitrarily selected based on the need to retain a certain percentage for breeding purposes, the mean of the selected individuals above the corresponding truncation point can be estimated based on the percentage being saved and the resulting value is referred to as the selection intensity. This approach, however, assumes the population is normally distributed. For practical purposes, especially with aquatic species that tend to produce skewed distributions for many traits, it is usually more accurate to determine a true mean for all selected individuals. Occasionally, distributions can be mathematically transformed to approximate normality, but these approaches can be difficult to apply under field conditions. In practice, selection intensity (i) is typically expressed in standard deviations (of the population mean) (Figs 4.2, 4.3 and 4.4). The genetic progress that can be expected in one generation will vary in relation to the intensity of selection that can be applied, and is roughly equal to $i \times \sigma P \times h^2$.

An examination of this formula reveals that progress can also be limited by the variation available to select from (σP), and by the degree of agreement between observed phenotypes and individual breeding values (h^2). Similarly, and perhaps more importantly for most aquacultured species, this formula can be used to derive an estimate of realized heritability, h^2. Realized heritability estimates can be developed from real-world selection trials based on observed phenotypic variances, σP, and selection intensity, i.

Clearly, in certain aquatic species much higher degrees of selection intensity can be applied to males than females (Fig. 4.5). This, however, is a double-edged sword, as it greatly increases levels of inbreeding from generation to generation (see Chapter 5). It is also apparent that when population sizes are being increased selection cannot be as

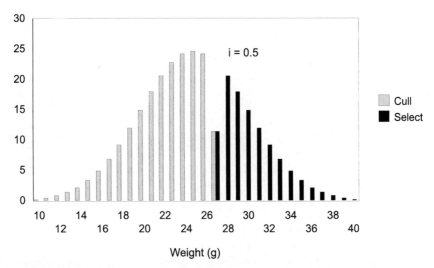

Fig. 4.2 Portion of population selected when i = 0.5 standard deviations.

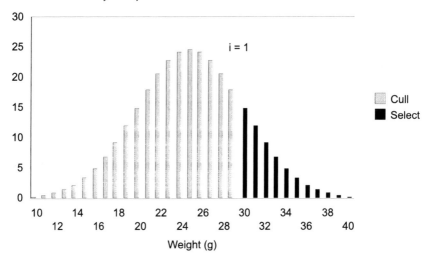

Fig. 4.3 Portion of population selected when $i = 1.0$ standard deviations.

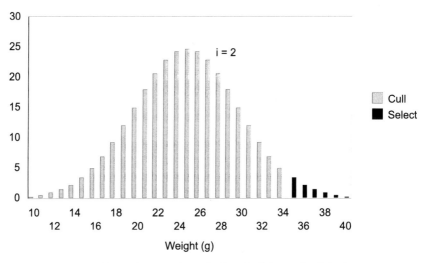

Fig. 4.4 Portion of population selected when $i = 2.0$ standard deviations.

intense as when numbers are held more or less constant from generation to generation. An additional consideration involves potential changes in gene frequencies, genetic variance, and consequently phenotypic variance, as selection proceeds from generation to generation. This variance need not always decline – in fact, it will be maximized at intermediate gene frequencies and so may actually increase for some period after selection has begun.

4.2.3 Correlated responses

Selection for one particular trait will often result in measurable changes in other traits as well. This is because many, if not most, gene prod-

Fig. 4.5 The fingerling stage may be the most practical phase of the production cycle in which to perform selection for many species. Studies indicate that early growth in channel catfish (*Ictalurus punctatus*) is highly correlated with age at marketable size. Care must be taken to ensure sufficient numbers of the slower-growing gender, in this case females, are retained for future breeding needs. (Photo courtesy J.L. Avery.)

ucts can be expected to influence more than one trait. The sum of these relationships for any two traits can be described and estimated as their genetic correlation. Possibly the most practical approach to developing such estimates is the use of a nested mating design, as described in Chapter 3. This type of analysis allows computation of genetic correlations between traits in addition to variances of individual traits. In terms of selection, only those genetic correlations relating to additive covariance need concern us from a practical standpoint, but all genetic correlations are generally difficult to estimate with any degree of precision. When attempting to alter any particular trait through selection it is wise simply to monitor other economically important traits to determine to what degree, if any, they are being improved or impacted through indirect selection. If negative correlations are suspected, a selection index can be developed to maximize gains in two or more traits (see below).

4.2.4 Multi-trait approaches

The concept of genetic correlations and indirect selection becomes a particularly important consideration in many aquatic species. Since the economic value of an animal may be largely determined by more than one trait, overall breeding goals may dictate progress in two or more traits simultaneously. Trade-offs are usually required in such a situation, however, since selection intensity is somewhat reduced in each trait included in the selection program. How much it will be reduced for any given trait will depend on genetic correlations between the traits, the number of traits involved, and the relative economic importance of each particular trait.

Three methods are available for improving two or more traits: tandem selection, independent culling, and development of a selection index. Tandem selection relies on first improving one trait, then selecting for a second trait within the "improved" population, and so on. The problem with this approach is a lack of progress altogether in those traits not being selected upon in any given generation. Additionally, the genetic variation for those traits selected for in subsequent generations may be substantially reduced. Independent culling, in contrast, involves establishing a minimum performance standard in each trait of interest. For example, a breeder of red tilapia hybrids might require a minimum weight of 400 grams at 7 months of age and a maximum surface area of roughly 5% mottling. If, say, 25% of the harvest were required for breeding stock, however, the maximum selection differential of 1.30 standard deviations attainable for either weight gain or mottling individually would be reduced to only 0.8 standard deviations for each of these (theoretically) independent traits. Progress per generation, in turn, would be reduced by roughly 38%.

In the selection index approach, a total score is computed for each animal, reflecting performance in each of a number of traits. Selection is then carried out based on the index score. In this way, animals with relatively unimpressive performance for one trait may still be selected based on outstanding performance in another. Genetic correlations between traits can be incorporated in index selection, but rarely will a correlation between two traits be known with sufficient precision to justify its incorporation. Nonetheless, index selection can result in real progress in many situations, even when genetic correlations are not directly accounted for.

In terms of overall genetic and economic improvement, if we assume traits are independent and have equal economic importance, heritabilities and variances, index selection is generally superior to independent culling. Under these assumptions, the expected progress for each trait when improving two equally important traits under index selection would be 71% of that possible for the trait if selected individually. For a three-trait index, the expected progress would drop to roughly 58% per trait, assuming equal importance, but this would still approximate that possible with two-trait independent culling.

Even when assumptions of equal economic value, heritability and variance are relaxed for two or more traits, the relative utility of these three approaches remains the same, at low or high selection intensities. When heritabilities are not known for the traits to be included in a selection index, the best weights to use are their relative economic values. These weightings can be determined, one trait at a time, by estimating the change in value that would occur (per individual) if the trait were increased by one unit, whatever the unit is that the trait is measured and recorded in. The result is an index that takes into account not only the phenotypic scores for individual traits, but the economic importance of each of those traits.

4.2.5 Complicating and constraining factors

Often, progress attained through selection will be less than what might have been predicted based on initial estimates of heritability, even when a control line is available for comparisons. A number of factors can be involved in these situations, including inbreeding depression, loss of epistatic effects upon relaxing selection pressure, random genetic drift, changes in fecundity within selected lines, unequal fecundity among individuals, unanticipated negative correlated responses in other economically important traits, genotype–environment interactions and frequent changes in breeding goals. A tendency prevails to consider heritability estimates as predictive, rather than descriptive. This, of course, is a fundamentally flawed interpretation of the estimate and its meaning in the strictest sense. Traits that are more closely associated with fitness are sometimes less likely to conform to the progress that would be expected under selection.

Alleles that impart superior performance in one production environment may not necessarily do so in other environments, necessitating field trials or selection under commercial conditions to verify the superiority of the stocks being developed (Fig. 4.6). Whenever possible, an unselected control line should be established in a selection trial and

Fig. 4.6 Variation in algal assemblages and resulting water quality inconsistencies can result in different environmental conditions from generation to generation or pond to pond during selection trials. (Photo courtesy USDA ASCS: public domain satellite image.)

maintained in the same manner as the selected line(s). This can provide an indication of changes due to domestication selection, inbreeding, or other factors. Additionally, establishment of both positive- and negative-selected lines will often provide valuable information from which to plan improvement programs.

4.2.6 Improving selection efficiency

One of the simplest means of improving the efficiency of selection is to improve the degree to which individual genetic values are estimated. This can be done in many ways, one of which involves compiling repeated measurements of performance in certain traits (such as fecundity, egg quality, sperm motility, etc.). Age trends must be accounted for when developing records based on repeated measurement on such reproductive traits. Another approach involves accumulating data on relatives of individual animals. In dairy-cattle breeding, early use of information from dams and sires of both cows and bulls allowed roughly twice as much genetic progress per generation as would have been possible by mass selection alone. Modern herd data management practices allow for the use of even more information from other types of relatives. Whether developing these types of information is actually worthwhile for most aquatic species, however, will depend on the extent to which selection must be delayed and generation intervals extended while data are compiled, as well as on the practicality of tracking performance of close relatives such as siblings, parents or progeny. In many aquatic species, these efforts cannot be justified in commercial production settings, but may be appropriate in hatchery operations specializing in seed production.

One direct approach to gathering information from an individual's relatives is progeny testing, or selecting individuals (at least partially) based on the performance of their offspring. Progeny testing, in spite of its inherent need to increase the generation interval, is useful when selecting for traits with low heritability, traits that are expressed in only one sex, or traits that describe slaughter-related characters such as dressout percentage. Progeny testing also lends itself well to many aquatic species since the required number of progeny can usually be produced within a reasonably short period. Although more information is generated for characterization and evaluation, the individual is still the unit being selected on.

Another approach to selection is to estimate an individual's genotype, wholly or partially, based on the performance of its siblings (Fig. 4.7). This type of correlation between relatives, however, cannot be larger than 0.5 if half-sibs are used or 0.71 if full-sibs provide the information. The progress per generation theoretically possible using information from half-sibs is only half that attainable using progeny performance as a basis for selection. The trade-off lies in the reduction of the generation interval when using information from siblings.

Fig. 4.7 Dressout percentage is an important production trait in many aquatic species, such as channel catfish (*Ictalurus punctatus*), pictured here. To select for dressout percentage, highly-correlated non-lethal measurements must be developed, or a form of family selection must be practiced where individuals are scored based on the performance (dressout percentage) of a sample of their siblings. (Photo courtesy J.L. Avery.)

4.2.7 Using family data

Selection can be based solely on family averages. This practice is known as family selection, where all (or a random sample of) members of those families with the highest averages are retained for breeding purposes. For traits with low heritabilities, family selection is generally more efficient than mass selection. When families include high numbers of siblings, as is the case for many aquacultured species, the efficiency of mass selection and family selection becomes roughly equal at heritabilities of 0.25 for half-sibs and 0.50 for full-sibs. An inability to identify aquatic individuals to particular families in many commercial settings, however, often requires growers to rely on mass selection even for traits with low heritabilities.

The judicious avoidance of sources of environmental or maternal bias for particular families, which might serve to reduce or inflate phenotypic variance and alter the correlation between phenotype and genotype, must always be considered a key aspect of refining the degree to which measured values reflect genetic values, whatever form of selection one wishes to practice. Using individuals' deviations from group averages has sometimes been proposed as a means to avoid bias from common environments and trends over time, but thought must be given as to which, if not all, of its contemporaries an individual will be evaluated against (half-siblings, full-siblings, all contemporaneous individuals within the same pond, or on the same farm, etc.) Common environments, and competition or behavioral hierarchies within environments become important considerations in these evaluations.

Nonetheless, in some circumstances the optimum selection approach involves ranking individuals within families. This approach, known as within-family selection, may be of particular value when families must be segregated in order to track their performance. Under some conditions, such as when broodstock are spawned en masse in groups, this approach is generally not possible. Even under the most similar conditions attainable, subtle differences in culture environments can still exist among family groups, but with this approach they should not directly bias the evaluation criterion. Within-family selection is most efficient when differences between families are largely due to differences in their rearing environments. An added benefit of this approach is the ease involved in selecting two members of every family to replace the preceding generation. This is an extremely valuable practice, because when every family contributes equally to the breeding population the effective population size theoretically becomes two times the actual population size. In this way, the accumulation of inbreeding depression is slowed considerably.

Alternately, an individual's own performance can be combined with that of its siblings to provide a composite score. Modeling and quantitative theory indicate that selection gains based on a combination of individual performance and family average will consistently surpass those of individual or family selection. Again, however, the requirement to identify and track each individual may not justify the additional efficiency of this type of selection.

4.3 PRACTICE

4.3.1 Implementation difficulties

Parent–offspring regression analysis can be applied to some aquaculture species more easily than others. In many species, tracking the parentage of individual animals in a common environment may not be particularly difficult. In others, however, it may be nearly impossible without imposing extremely artificial conditions. Calculation of parent–offspring regressions requires at least two generations, which can become a long-term proposition for many aquatic species such as channel catfish, barramundi, red drum or sturgeon, to name just a few.

An additional sticking point to the widespread use of parent–offspring regression in aquaculture is the assumption that all animals, offspring and parents, are raised under the same environmental conditions. Within each generation, water quality, temperature, stocking density and the amount, quality and delivery of feed become important considerations when estimating parent–offspring regressions. Influences well beyond the obvious may come into play. For example, research results have suggested that diet can have a significant effect on cold tolerance in red drum (*Sciaenops ocellatus*). If ration formulation (or any of a number of other variables) cannot be held constant across an entire selection study, the potential for confounding effects in a parent–offspring regression is readily apparent.

Clearly, the assumption of constant environment across generations may not always be reasonable depending on the trait in question or the available facilities. Accordingly, an unselected control line should always be available to serve as a reference for changes due to rearing environment, or domestication selection (see below). The need to maintain and evaluate one or more control lines clearly adds to the expense of a selection trial, but the information generated is usually far more valuable than if no control were available. As in single-generation analyses, the question of how well the experimental environment reflects conditions of commercial culture also becomes an issue. Finally, if the variance of parental scores or values is substantially different than that of the progeny, statistical problems can also arise.

One unavoidable by-product of many selection programs is inbreeding. Although this topic will be discussed in more detail in Chapter 5, suffice it to say that limiting reproduction to a relatively small segment of the population selected for particular phenotypes will, over time, result in mating of more and more closely related individuals. Strategies can be developed to reduce the accumulation (and resultant impacts) of inbreeding, but no form of selection can completely avoid this problem.

4.3.2 Identification options

Use of marking techniques such as cold- or heat-branding, fin clipping and other approaches can assist with the execution of mating programs that limit mating between closely related broodstock. These techniques can occasionally facilitate the collection of data from relatives, allowing for more precise estimation of the breeding values of selected individuals. A simple heat-branding device can be constructed using 20-gauge nickel-chrome wire attached to a 6-volt battery. Care must be taken to avoid injury and ensure brands are applied superficially.

One innovative technology that is becoming more practical to acquire and use relies on internal tags, known as PIT (passive integrated transponder) tags. A portable electronic device can read these tags while they are still within individual fish, allowing for easy identification of individuals without the need for external markings. In group-spawning fish such as seabream (*Sparus aurata*), these techniques are of little use since the parents of any particular fish cannot be readily determined. While molecular techniques are available that can often identify direct genetic links between individual offspring and broodstock, they are entirely impractical under most commercial or artisanal conditions. The result is that inbreeding becomes more difficult to control in group-spawning species.

4.3.3 Lack of response to selection

Consider an example that ties in the concepts of inbreeding and selection response. What would happen if a hatchery manager spent considerable time and money carefully selecting and maintaining the biggest, best broodstock produced at his facility only to find over the next 12 months the offspring they produced were run-of-the-mill? If this had occurred with just one or a few phenotypically superior individuals whose progeny could be identified directly, we could assume their positive attributes were mainly due to lucky environmental circumstances over the course of their lifetimes.

On the other hand, if this disappointment was the rule across the board for all the selected broodstock on hand, our manager would have to accept the harsh fact that breeding values and additive genetic variance played little if any role in the expression of the trait(s) he wanted to improve, at least in the population he was working with and under the conditions of his particular operation. In short, there would be little point in any further pursuit of selection.

This is not an uncommon situation, and it can often be the result of limited genetic variation within the population in question. Even if a trait is highly heritable, that is, highly influenced by additive genetic effects, in the absence of variation to work with there can be little progress from generation to generation. If the facility were growing, say, Nile tilapia or freshwater prawns somewhere in North America, the animals in question might well already be the products of many generations of inbreeding of close relatives from a relatively small founding population originally collected from wild stocks. Similar circumstances might apply to tilapia stocks in various locations, or stocks of channel catfish in countries such as China. Getting "new blood" from other sources may help in these situations. If we are concerned with a trait controlled largely by additive effects, however, the "new blood" in question must be genetically superior. With additively controlled or additively influenced traits, more distinct does not necessarily mean more desirable. Only better is better when additive genetic effects are being considered.

4.4 ILLUSTRATIVE INVESTIGATIONS AND APPLICATIONS

4.4.1 Evaluating available strains

Inasmuch as they represent isolated breeding populations with their own distinct genetic attributes, the first step of selection in many aquaculture settings is to characterize available strains or lines within a species and subsequently select one or more that excel in certain areas, from which to form a base population for subsequent improvement. Eknath et al. (1993) reported (by consensus, one might hope) on the growth performance of eight strains of Nile tilapia (*Oreochromis niloticus*) under different commercial farm environments in the Philippines. As a foundation for the GIFT program (see below) four strains were from the Philippines and four were recently translocated from Africa, namely Egypt, Ghana, Kenya and Senegal. Highly significant differences were determined in growth among these strains in farm environments, and three of the African strains (all but Ghana) performed comparably to or exceeded the growth of the local Philippine strains. In outdoor concrete tanks, however, Bolivar *et al.* (1993) reported that seven of these eight strains (the Kenya strain was not evaluated) showed no differences in growth, with the exception of the Ghana strain, which exhibited a significantly lower body weight after 210 days of growout.

Notable differences in growth among strains of *Mugil cephalus* were reported by Tamaru *et al.* (1995). In response to local concerns among mullet producers in Guam, a series of studies was conducted to compare the growth of a hatchery strain produced in Hawaii and wild-caught fingerlings from two strains found in Taiwanese waters. Growth rate in captivity was three to four times greater for the wild mullet strains. These wild strains also exhibited much lower proportions of males as well as greatly reduced maturation levels at the end of the study.

4.2.2 Domestication selection

Occasionally, the processes of domestication and selection may occur simultaneously. Domestication itself can be interpreted as a form of selection, in which only those wild individuals capable of surviving the stress and artificial conditions of a culture environment contribute to lines of captive-bred offspring. Doyle (1983) addressed this topic in detail, with an emphasis on domestication processes in aquatic species. Hershberger *et al.* (1990) reported on the results of 10 years of selection for growth of coho salmon (*Oncorhynchus kisutch*) in marine net-pens. In the odd- and even-year broodstocks developed during the study, realized heritability estimates for growth were roughly four to six times higher than those determined by analysis of variance. The authors explained these discrepancies as the result of a combination of direct selection and domestication selection, since internal control lines exhibited a comparable response over time, while wild controls remained relatively unchanged over the course of the study.

These hypotheses are supported by previous findings in channel catfish (*Ictalurus punctatus*). Dunham and Smitherman (1983) compared responses to selection for body weight and realized heritabilities in three strains of channel catfish in earthen ponds. Fish with shorter periods of domestication generally exhibited greater responses to selection.

4.4.3 Conflicting results

Working with a population of Nile tilapia (*Oreochromis niloticus*), Huang and Liao (1990) determined that "mass selection is not an effective means of improving growth rate in this species." Perhaps a more prudent interpretation would have been restricted to the strain in question. Since heritability is merely a description of the relative amount of additive variance within a population, highly inbred lines tend to exhibit lower heritabilities, as a general rule. The fish in the Huang and Liao study were produced from two lines that had been maintained in laboratory tanks for seven years. Prior to this, both lines had been propagated at commercial farms from offspring of a total of 56 individuals, originally introduced from Japan 20 years earlier. No mention was made as to whether these 56 individuals were from a single spawn.

Similarly, Hulata *et al.* (1986) had previously reported that no improvement was achieved by mass selection for growth in Nile tilapia (Fig. 4.8). Their "original parent" control group performed as well or better than progeny of fish selected at adult or juvenile stages, respectively. All fish in the study, however, were descended from a sample of offspring from a stock established seven years earlier consisting of only "around 50 fingerlings" introduced from Ghana. Again, no reference was made as to whether the founding fingerlings were from more than one full-sib family.

Fig. 4.8 Although selection for growth has yielded mixed results in Nile tilapia (*Oreochromis niloticus*), substantial gains have been achieved when selection was imposed on a synthetic crossbred base population developed from multiple strains and subspecies. (Photo courtesy J. Hargreaves.)

However, other researchers working with this species were not dissuaded from pursuing selection as a means to improve production stocks. From 1988 through 1997, the GIFT (Genetic Improvement of Farmed Tilapia) project was undertaken in the Philippines. A joint effort by several agencies, the project involved the development of a synthetic base population including eight strains of Nile tilapia, four from the Philippines and four imported directly from Africa (Eknath et al. 1993). Combined family and within-family selection for growth was applied on an annual basis across five generations. The average selection response per generation was 13% and the accumulated response resulted in an 85% increase in growth (Gjedrem 1999).

Even when using heritability estimates within a population, as opposed to making blanket predictions for an entire species, results may not always be in line with expectations. Hoerstgen-Schwark (1993) indicated that selection responses for growth in rainbow trout (*Oncorhynchus mykiss*) were notably higher in low density culture than at high densities. Crenshaw et al. (1996) reported on conflicting realized heritability estimates for growth within a population of *Mercenaria mercenaria*. Under moderate densities (roughly 1 cm^{-1}) realized heritability for growth was calculated to be 0.402. At high densities (roughly 4 cm^{-1}) however, control progeny grew significantly faster than select progeny, resulting in negative realized heritability.

In previously reported work on realized heritability for growth in *Mercenaria mercenaria* Hadley et al. (1991) described consistent results in two selection trials. Selection was applied to clams at 2 years of age, with the largest 10% of the population and an equal number of clams approximating the population mean value selected to produce select and control lines, respectively. This approach to forming a control population involves what might be considered assortative mating, which may not be applicable in some situations but generally is of little consequence in parent–offspring regressions and may actually serve to improve the precision of the realized heritability estimate. Progeny from the two lines were measured at 2 years of age and realized heritabilities computed. In two trials, heritability estimates were in good agreement: 0.42 ± 0.10, and 0.43 ± 0.06. In another trial, however, no response to selection was observed. The authors speculated that effective breeding numbers within lines in this trial may have been so low that no response could be determined.

4.4.4 Correlated responses

Some correlated responses are generally anticipated when selecting for specific traits in aquatic species. Toro (1990) addressed genetic correlations between live weight and shell height in the European oyster *Ostrea edulis*. Individual oysters were selected from each of 40 full-sib families, based on their whole live weight at the end of one growing season. Six downward-selected and 18 upward-selected families were produced, resulting in 6560 individually tagged spat. After the first and second growing seasons, live weight and shell height were recorded in the field for each oyster and heritability estimates were developed for each trait. Heritabilities for live weight and shell height ranged from 0.112 to 0.243, and genetic correlations between the two traits were estimated at 0.963 and 0.995 after the first and second seasons, respectively.

Other potentially correlated responses may be less obvious before selection is undertaken. In many strains of channel catfish (*Ictalurus punctatus*) selected for rapid growth, the onset of maturation and reproductive activities is delayed. In some cases, this has caused certain select strains to be unsuitable for use as commercial breeding stock, since by the time they begin to spawn they are too large to use conventional spawning refuges (typically steel boxes or cans submerged in brood ponds) and too difficult to handle (at weights of 3 to 5 kg each). Costs of maintaining these fish are often difficult to justify from a hatchery operator's standpoint. When comparing select and control lines of the Kansas strain of channel catfish, Dunham and Smitherman (1982) reported that although no statistically significant differences were found in spawning day, spawning rate, hatchability of eggs or survival of sac fry, total fingerling output was significantly lower for the select line. In spite of the fact that fecundity was higher in the select line (though not significantly), lower spawning rates, lower hatchability, and reduced fry survival combined to result in the observed reduction in fingerling output.

4.4.5 Indirect selection through production practices

Doyle *et al.* (1983) addressed the problem of indirect selection in aquaculture by examining commercial practices and computer-simulated populations of freshwater prawns (*Macrobrachium rosenbergii*). In both real and simulated populations, they demonstrated that taking the required female broodstock for farm operations from pond populations in early harvests (say, at 3 months post-stocking) would result in indirect selection for improved growth. In contrast, collecting female broodstock from late harvests (6 months post-stocking or later) would result in negative selection for growth. These concerns have also been raised for other crustaceans selectively harvested by retention in trapping or seining (Fig. 4.9). Similarly, based on results of a study of the genetics of weight and age at sexual maturity in rainbow trout, Crandell and Gall (1993a) concluded that selection to decrease the age at maturity in this species could affect body weight at maturity and associated traits relating to egg production and quality.

4.4.6 Indirect measurement

Indirect selection need not be inadvertent. Dunham *et al.* (1985) examined the potential for improving dressout percentage in channel catfish (*Ictalurus punctatus*) through indirect selection on a variety of measurements, many of which could be taken on living fish without requiring their slaughter. Traits such as length and width of the head,

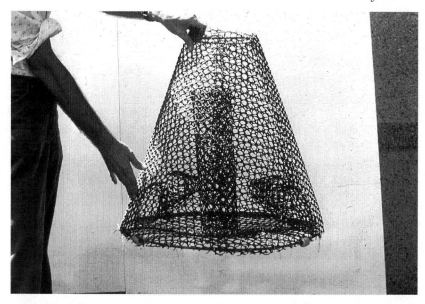

Fig. 4.9 Frequently, methods of harvest that rely on retention of animals above a minimum size, such as this crayfish trap, may inadvertently cull faster-growing individuals and select for slow growth or reduced size at maturity. (Photo courtesy R.P. Romaire.)

body depth, girth, specific gravity and body density were all evaluated as non-lethal measures to predict dressout percentage. Of these, body density was the most promising, with phenotypic correlations of −0.64 and −0.61 with dressout percentage for females and males, respectively.

Lester (1983) examined the relative merits of various approaches to selective breeding for penaeid shrimp mariculture and concluded that, under commercial methodology utilized in Texas at that time, mass selection would be the most efficient approach toward genetic improvement of tail weight. After examining correlation coefficients among ten morphological variables in *Penaeus stylirostris* and *P. vannamei*, Lester suggested that measurement of the depth of the sixth abdominal segment in these species could provide a suitable criterion on which to select for tail weight within a breeding population. Correlation coefficients of this measurement with tail weight were 0.95 and 0.85 for *P. vannamei* and *P. stylirostris*, respectively. Additional advantages to using this particular measurement included ease of acquisition, relative precision due to resistance of this segment to compression and the fact that injury to shrimp would be unlikely during required handling and data collection associated with use of dial calipers.

4.4.7 Altering environmental tolerances

Wong and McAndrew (1990) reported on progress in efforts to develop a totally freshwater strain of *Macrobrachium nipponense* with freshwater-tolerant larvae. Over three generations, larval survival in fresh water was increased by over 50% in two families. The average heritability for larval tolerance of fresh water over all generations and families in the study was calculated as 0.24 ± 0.065. Cheong (1986) related results of selection for cold tolerance in *Macrobrachium rosenbergii*. After one generation of selection for cold tolerance in two separate select lines, responses in lethal cold temperature were 0.3°C in one line and ranged from 0.2–0.6°C in the other. Pooled data resulted in realized heritabilities of 0.5 ± 0.04 and 0.9 ± 0.04 for the two select lines. Correlations between growth and lethal cold temperature were not significant. Many similar efforts to alter environmental tolerances have proven successful with a variety of aquatic species.

4.4.8 Adjusting data for environmental bias

Tave (1994) presented results of selection trials on the golden shiner, *Notemigonus crysoleucas*. An initial analysis of data suggested a significant difference between control and select lines after one generation of selection for growth in length. After statistical adjustment of the data for density effects, however, differences were still apparent but not statistically significant. Realized heritability for adjusted data was estimated to be 0.42, suggesting at least some potential benefit in selecting for growth rate in this species.

4.4.9 Accounting for differences between sexes

Crandell and Gall (1993b) addressed an important topic which is often overlooked or avoided in selection trials: the effects of sex on traits such as growth. They estimated heritabilities for body weight of individually tagged rainbow trout (*Oncorhynchus mykiss*) by taking ten weight measurements on each fish between 159 and 740 days post-fertilization. Two data sets were developed: one with records for all fish, including some for which sex was not determined and another with records only for fish of known sex. Sex, as might be expected for this particular species (see Chapter 8), had a statistically significant influence on body weight. More importantly, however, sex influenced heritability estimates. For weights measured at 362 days and later, heritability estimates derived with a model that included a fixed effect for sex were over 50% larger (0.08 to 0.20 points), than those developed with a model that ignored sex.

4.4.10 Genotype by environment interactions

Occasionally, the additive genetic attributes that result in superiority in one environment may confer no advantage under other circumstances. Romana-Eguia and Doyle (1992) illustrated this point in an examination of the performance of three strains of tilapia (*Oreochromis niloticus*) under conditions of alternating adequate and poor nutrition. When fish were fed a commercial feed during the first and last two weeks of a six-week period, the NIFI strain grew best. During the middle two weeks, fish were fed rice bran, and in this period the "Israel" strain excelled significantly in weight gain. A third Philippine strain, known as CLSU, ranked last in all phases of the growth trials.

4.4.11 Miscellaneous results: finfish

Occasionally, selection may be undertaken with the objective of altering morphological or meristic characters. Although this approach is generally more common in ornamental fish culture, it is also occasionally a goal in species grown for human consumption. Ankorion et al. (1992) presented results of bidirectional mass selection for body shape in common carp (*Cyprinus carpio*). The selection criterion was the ratio of body height to length. A realized heritability estimate of 0.47 ± 0.06 was determined for increasing body depth, while an estimate of 0.33 ± 0.10 was calculated for reducing body depth. The overall realized heritability for a change in either direction, based on the divergence of the two lines, was reported as 0.42 ± 0.03.

Working in an extremely practical setting, Brzeski and Doyle (1995) examined the results of mass selection for size-specific growth rate of tilapia in a commercial fish farm in Indonesia. Selection was applied in two steps over a 6-month growout period, and initial stocking was based on size-graded, rather than completely contemporaneous, fingerlings. These procedures reflected the practical needs of small-scale non-technical producers in the region (and elsewhere) (Fig. 4.10). Offspring of select and control lines were evaluated in hapas. Although the selection response was small (a 2.3% improvement in weight gain) it was statistically significant, and might be expected to accumulate appreciably over time based on the relatively short generation interval in tilapia. Realized heritability was estimated to be 0.12.

Wiegertjes et al. (1994) examined a novel approach to selection for antibody production in common carp (*Cyprinus carpio* L.). From a base population of 101 carp, they selected broodfish based on early/high or late/low antibody responses. Three females from each group were spawned through gynogenesis (see Chapter 6), resulting in homozygous progeny groups. These progeny were then evaluated for their antibody responses, and although all groups responded more slowly than the base population, presumably as a result of high levels of inbreeding (see Chapter 5), significant differences were noted between high- and low-selected groups. Realized heritability for antibody production was calculated to be 0.37 ± 0.36.

Knibb (2000) provided a balanced overview of genetic improvement approaches for gilthead seabream (*Sparus aurata*). Since this species is normally induced to spawn in large groups under maturation conditioning (see Chapter 9), families cannot be readily identified, leaving mass selection as the only available selection approach in a commercial setting. Under these constraints, selection for growth in weight still resulted in gains of 5–10% per generation.

88   *Practical Genetics for Aquaculture*

Fig. 4.10   Perhaps one of the most practical approaches to selection for growth in finfish involves mechanical grading by size. (Photo by the author.)

4.4.12   Miscellaneous results: mollusks

Much interest has been expressed in culture of the southern bay scallop, *Argopecten irradians concentricus*, in the southeastern United States. Crenshaw *et al.* (1991) reported on the realized heritability of growth in this species. Using a population derived from wild stocks in Florida and Georgia, they selected a random control line prior to selecting the largest 15.9% of the population. Based on performance of offspring produced by the two groups, realized heritability in this scallop population was estimated to be 0.206.

When replicate selected lines can be accommodated by available space and labor, not to mention a comparatively cooperative species that lends itself to induced spawning, results are usually much more instructive and reliable. Working with the native Chilean oyster (*Ostrea chilensis*) Toro *et al.* (1994) produced five subgroups each of high-selected animals, low-selected animals and unselected controls for live weight at 40 months of age. Realized heritability ranged from 0.43 ± 0.18

to 0.69 ± 0.11 for positive selection, while estimates for negative selection ranged from 0.24 ± 0.06 to 0.35 ± 0.08.

The commercially important oyster *Saccostrea cucullata* was the focus of an impressive study in realized heritability reported on by Jarayabhand and Thavornyutikarn (1995). A base population of oysters on which to conduct selection was reared at a demonstration farm for 15 months, at which time positive- and negative-selected groups and a control line were established based on whole weights. These oysters, in turn, were spawned to produce progeny for evaluation. When the majority of these progeny had attained marketable size, whole weight means for the three groups differed significantly. The overall realized heritability for growth rate was estimated to be 0.277 ± 0.006, indicating good potential for growth improvement (at least in the base population) through mass selection.

Another oyster in the genus *Saccostrea*, the Sydney rock oyster (*S. glomerata*) has been found to respond to selection for growth. Nell *et al.* (2000) outlined ongoing breeding programs for this species in Australia. Following two generations of mass selection for weight gain, breeding lines exhibited 14–23% improvement (18% on average). The authors equated this gain with a 3-month reduction in growout to marketable size. Problems with protistan parasites complicated selection trials, but preliminary results suggested gains in parasite resistance had also been attained.

4.4.13 Miscellaneous results: crustaceans

Henryon (1996) summarized a number of studies relating to the potential for improving commercial strains of marron, the large Australian crayfish (*Cherax tenuimanus*). Marron from four different river systems exhibited notable differences in breeding success, as well as growth and tail yield. The latter two traits appeared to be negatively phenotypically correlated. Heritability estimates for growth, tail size and chelae size ranged from 0.3 to 0.6. Economic projections suggested the use of one particularly fast-growing strain would result in an 80% increase in profit over that which would result when utilizing a slow-growing population. Crosses between four strains did not result in measurable heterosis for growth, tail yield or chelae size.

In response to the tremendous economic impact of shrimp farming in many regions of the globe, a number of selection studies have been conducted with penaeid shrimp in recent years. Goyard *et al.* (1999) reported that selected lines of *Penaeus stylirostris* demonstrated a 21% increase in growth after five generations of mass selection, when compared to a non-selected control line. Crocos *et al.* (1999) indicated gains of 10–15% in weight at first harvest for their selected Australian lines of *Penaeus japonicus*. Moss *et al.* (1999) reported mean heritabilities of 0.40 ± 0.06 for weight gain and 0.09 ± 0.03 for TSV resistance in *L. vannamei*. Wyban (1999) described selection results for resistance to Taura Syndrome Virus (TSV) in the commercially produced shrimp *Litopenaeus vannamei*. After three rounds of selection for viral resistance, select lines averaged 69% survival following a TSV challenge, compared to 31% for an unselected control line.

The redclaw crayfish *Cherax quadricarinatus* has been widely introduced throughout the globe from its native northern Australia. Jones *et al.* (2000) reviewed recent efforts in genetic improvement in this species. A comparison of several strains of redclaw suggested that size at maturity varied substantially within strains but not between them. As has been described for other crayfish, mature females were found to exhibit significantly larger abdomens than males, while chelae of males were significantly larger than those of females. Selection for growth in two distinct strains resulted in significant differences between strains, sexes and select and control lines. Based on a selection differential of 19.4 ± 1.1 g and a response of 4.7 ± 1.9 g, realized heritability was calculated to be 0.24 ± 0.06. One of several problems in the trials was the escape of some stock from pond pens, potentially resulting in some degree of mixing between strains and/or lines. Nonetheless, since this would be expected to downwardly bias the resultant heritability estimate, the analysis suggests good potential for improvement of redclaw growth through selection.

The taxonomic status of the Australian crayfish known collectively as yabbies has been the subject of numerous disputes over the years. Three main divisions within this "yabby complex" are referred to as the destructor, punctatus and dispar groups. Within these groups, a number of isolated species, subspecies and strains exist (Lawrence & Morrissy 2000). The yabby complex as a whole includes populations adapted to a great variety of environmental conditions. Lawrence and Morrissy reviewed recent efforts to development genetically improved yabby stocks for culture. They documented growth differences among a number of yabby strains collected from different geographical regions. The fastest growing strain exhibited a daily gain roughly nine times higher than that of the slowest growing strain.

4.5   REFERENCES

Ankorion, Y., Moav, R. & Wohlfarth, G.W. (1992) Bidirectional mass selection for body shape in common carp. *Genetics Selection and Evolution*, **24** (Suppl. 1), 43–52.

Bolivar, R.B., Eknath, A.E., Bolivar, H.L. & Abella, T.A. (1993) Growth and reproduction of individually tagged Nile tilapia (*Oreochromis niloticus*) of different strains. *Aquaculture*, **111**, 159–169.

Brzeski, V.J. & Doyle, R.W. (1995) A test of an on-farm selection procedure for tilapia growth in Indonesia. *Aquaculture*, **137** (Suppl. 1–4), 219–230.

Cheong, R.M.T. (1986) Response to selection and realized heritability for cold tolerance in juvenile *Macrobrachium rosenbergii* (DeMan). MS Thesis, Louisiana State University.

Crandell, P.A. & Gall, G.A.E. (1993a) The genetics of age and weight at sexual maturity based on individually tagged rainbow trout (*Oncorhynchus mykiss*). *Aquaculture*, **117** (Suppl. 1–2), 95–105.

Crandell, P.A. & Gall, G.A.E. (1993b) The effect of sex on heritability estimates of body weight determined from data on individually tagged rainbow trout (*Oncorhynchus mykiss*). *Aquaculture*, **113** (Suppl. 1–2), 47–55.

Crenshaw, J.W., Heffernan, P.B. & Walker, R.L. (1991) Heritability of growth rate in the southern bay scallop, *Argopecten irradians concentricus* (Say, 1822). *Journal of Shellfish Research*, **10** (Suppl. 1), 55–63.

Crenshaw, J.W., Heffernan, P.B. & Walker, R.L. (1996) Effect of growout density on heritability of growth rate in the northern quahog, *Mercenaria mercenaria* (Linnaeus, 1758). *Journal of Shellfish Research*, **15** (Suppl. 2), 341–344.

Crocos, P., Preston, N. & Lehnert, S. (1999) Genetic improvement of farmed prawns in Australia. *The Advocate, Global Aquaculture Alliance*, **2** (Suppl. 6), 62–63.

Doyle, R.W. (1983) An approach to the quantitative analysis of domestication selection in aquaculture. *Aquaculture*, **33**, 167–185.

Doyle, R.W., Singholka, S. & New, M.B. (1983) "Indirect selection" for genetic change: a quantitative analysis illustrated with *Macrobrachium rosenbergii*. *Aquaculture*, **30**, 237–247.

Dunham, R.A. & Smitherman, R.O. (1982) Effects of selecting for growth rate on reproductive performance in channel catfish. *Proceedings of the Annual Conference of the Southeastern Association of Fish and Wildlife Agencies*, **36**, 182–189.

Dunham, R.A. & Smitherman, R.O. (1983) Response to selection and realized heritability of body weight in three strains of channel catfish, *Ictalurus punctatus*, grown in earthen ponds. *Aquaculture*, **33**, 89–96.

Dunham, R.A., Joyce, J.A., Bondari, K. & Malvestuto, S.P. (1985) Evaluation of body conformation, composition and density as traits for indirect selection for dress-out percentage of channel catfish. *The Progressive Fish-Culturist*, **47** (Suppl. 3), 169–175.

Eknath, A.E., Tayamen, M.M., Palada-de-Vera, M.S. *et al.* (1993) Genetic improvement of farmed tilapias: The growth performance of eight strains of *Oreochromis niloticus* tested in different farm environments. *Aquaculture*, **111**, 171–188.

Gjedrem, T. (1999) Aquaculture needs genetically improved animals. *The Advocate, Global Aquaculture Alliance*, **2** (Suppl. 6), 69–70.

Goyard, E., Patrois, J., Peignon, J., Vanaa, V., Dufour, R. & Bedier, E. (1999) IFREMER's shrimp genetics program. *The Advocate, Global Aquaculture Alliance*, **2** (Suppl. 6), 26–28.

Hadley, N.H., Dillon, R.T. & Manzi, J.J. (1991) Realized heritability of growth rate in the hard clam *Mercenaria mercenaria*. *Aquaculture*, **93** (Suppl. 2), 109–119.

Henryon, M. (1996) Genetic variation in wild marron can be used to develop an improved commercial strain. PhD Thesis, University of Western Australia. [Cited in Lawrence, C.S. & Morrissy, N.M. (2000). Genetic improvement of marron *Cherax tenuimanus* Smith and yabbies *Cherax* spp. in Western Australia. *Aquaculture Research*, **31** (Suppl. 1), 69–82.]

Hershberger, W.K., Myers, J.M., Iwamoto, R.N., Macauley, W.C. & Saxton, A.M. (1990) Genetic changes in the growth of coho salmon (*Oncorhynchus kisutch*) in marine net-pens, produced by ten years of selection. *Aquaculture*, **85**, 187–197.

Hoerstgen-Schwark, G. (1993) Selection experiments for improving "pan-size" body weight of rainbow trout (*Oncorhynchus mykiss*). *Aquaculture*, **112** (Suppl. 1), 13–24.

Huang, C.-M. & Liao, I.-C. (1990) Response to mass selection for growth rate in *Oreochromis niloticus*. *Aquaculture*, **85**, 199–205.

Hulata, G., Wohlfarth, G.W. & Halevy, A. (1986) Mass selection for growth rate in the Nile tilapia (*Oreochromis niloticus*). *Aquaculture*, **57**, 77–184.

Jarayabhand,, P. & Thavornyutikarn M. (1995) Realized heritability estimation on growth rate of oyster, *Saccostrea cucullata* Born, 1778. *Aquaculture*, **138** (Suppl. 1–4), 111–118.

Jones, C.M., McPhee, C.P. & Ruscoe, I.M. (2000) A review of genetic improvement in growth rate in redclaw crayfish *Cherax quadricarinatus* (von Martens) (Decapoda:Parastacidae). *Aquaculture Research*, **31** (Suppl. 1), 61–67.

Knibb, W. (2000) Genetic improvement of marine fish – which method for industry? *Aquaculture Research*, **31** (Suppl. 1), 11–23.

Lawrence, C.S. & Morrissy, N.M. (2000) Genetic improvement of marron *Cherax tenuimanus* Smith and yabbies *Cherax* spp. in Western Australia. *Aquaculture Research*, **31** (Suppl. 1), 69–82.

Lester, L.J. (1983) Developing a selective breeding program for penaeid shrimp mariculture. *Aquaculture*, **33**, 41–50.

Moss, S.M., Argue, B.J. & Arce, S.M. (1999) Genetic improvement of the Pacific white shrimp, *Litopenaeus vannamei*. *The Advocate, Global Aquaculture Alliance*, **2** (Suppl. 6), 41–43.

Nell, J.A., Smith, I.R. & McPhee, C.C. (2000) The Sydney rock oyster *Saccostrea glomerata* (Gould 1850) breeding programme: progress and goals. *Aquaculture Research*, **31** (Suppl. 1), 45–49.

Romana-Eguia, M.R.R. & Doyle, R.W. (1992) Genotype–environment interaction in the response of three strains of Nile tilapia to poor nutrition. *Aquaculture*, **108** (Suppl. 1–2), 1–12.

Tamaru, C.S., Carlstrom-Trick, C. & FitzGerald, W.J. (1995) Differences in growth among strains of *Mugil cephalus*. *Proceedings, Sustainable Aquaculture '95*, Pacific Congress on Marine Science and Technology – PACON International, Honolulu.

Tave, D. (1994) Response to selection and realized heritability for length in golden shiner, *Notemigonus crysoleucas*. *Journal of Applied Aquaculture*, **4** (Suppl. 4), 55–64.

Toro, J.E. (1990) Respuesta a la seleccion, heredabilidades y correlacion genetica para los caracteres peso vivo y longitud de la concha en la ostra Europea *Ostrea edulis* Linne. *Revista de Biologia Marina*, **25** (Suppl. 2), 135–146.

Toro, J.E., Aguila, P., Vergara, A.M. & Newkirk, G.F. (1994) Realized heritability estimates for growth from data on tagged Chilean native oyster (*Ostrea chilensis*). *World Aquaculture*, **25** (Suppl. 2), 29–30.

Wiegertjes, G.F., Stet, R.J.M. & Van Muiswinkel, W.B. (1994) Divergent selection for antibody production in common carp (*Cyprinus carpio* L.) using gynogenesis. *Animal Genetics*, **125** (Suppl. 4), 251–257.

Wong, J.T.Y. & McAndrew, B.J. (1990) Selection of larval freshwater tolerance in *Macrobrachium nipponense* (de Haan). *Aquaculture*, **88** (Suppl. 2), 151–156.

Wyban, J. (1999) Selective breeding for TSV-resistant shrimp. *The Advocate, Global Aquaculture Alliance*, **2** (Suppl. 6), 30.

Chapter 5
Inbreeding, Crossbreeding and Hybridization

5.1 INTRODUCTION

Most of the genetic variation available for improvement of aquaculture stocks can be found in two distinct forms: additive variance, which we have already discussed in some detail, and dominance variance, which we will explore in this chapter. Dominance effects and the genetic variance they produce in metric traits are manifested in the familiar phenomena of hybrid vigor (also referred to as heterosis) and inbreeding depression.

Inbreeding depression, which is frequently associated with a general decline in fitness, can often be reduced from a practical standpoint simply by making special efforts to prevent closely related individuals from breeding with each other. It can sometimes be sufficiently offset by occasionally introducing new, unrelated individuals into a breeding program, if care is taken to obtain high-performing stocks with proper quarantine protocols. The practice of crossbreeding, or mating comparatively unrelated lines within a species, is also founded in dominance effects and their manifestation through heterosis. Although analyses used to estimate heritability can occasionally also provide some indication of the extent of dominance variance within a population and the potential for improvement through cross-breeding, the very nature of dominance effects often requires a try-it-and-see approach when developing breeding programs.

Improvements in production traits resulting from heterosis are not limited to crosses within species, but can often be attained through hybridization, the crossing of distinct (albeit closely related) species.

Apart from capturing the benefits of heterosis, hybridization is sometimes practiced simply to combine specific desirable attributes of related species. Another frequent goal of such efforts is the production of functionally sterile stocks, either for commercial culture or direct introduction into natural environments. A number of practical studies have been undertaken with aquatic species in pursuit of each of these goals.

5.2 THEORY

5.2.1 Dominance effects and multi-locus traits

If we regard the basis of additive genetic variance as the directional "scores" of the alleles present at each locus influencing a trait we are interested in, dominance effects might be conceptualized as accessory codes for each locus, such that certain code combinations increase or decrease the genetic value from the simple sum of the allelic values. As previously illustrated, dominance effects represent interactions between pairs of alleles at the same locus. Since these interactions depend on the specific alleles involved at that particular locus, they cease to exist during the formation of gametes through meiosis and are re-established anew in each generation as gametes combine and alleles re-pair randomly.

In quantitatively measured traits, influences of dominance deviations across all involved loci are reflected in dominance genetic variance. Apart from simple variance, these deviations can also influence the overall means of various traits as they manifest themselves both through heterosis and inbreeding depression. The expression of heterosis, where offspring of comparatively unrelated individuals exhibit increased fitness, is the basis for most hybridization and crossbreeding efforts. Inbreeding depression, in contrast, describes a general decline in fitness resulting from crossing closely related individuals. Since heterosis and inbreeding depression are both manifestations of dominance genetic effects, any discussion of one must usually refer to the other as well.

Perhaps the most immediate concern relating to dominance genetic effects that aquaculture producers around the world share is the ever-present suspicion that their production stocks are, or may become, "inbred." Often, this is a reasonable concern indeed. Inbreeding can be defined simply as the mating of two individuals whose degree of relationship is higher than the expected, or average, value would be if members of the population in question were allowed (or forced) to mate randomly. While many quantitative genetics texts address the calculation of inbreeding coefficients for individual breeding animals, this approach often has little use in commercial production of aquatic species. Frequently, the required pedigree information for individual fish, crustaceans or shellfish is unavailable or entirely impractical to generate. For most purposes, inbreeding need only be tracked over time for the breeding population in question.

A practical way to look at inbreeding is by determining the probability that both alleles at a given locus are identical by descent from a common ancestor. Figure 5.1 illustrates a simple approach to the evaluation of levels of inbreeding, known as an inbreeding coefficient. There is a 50% chance of two offspring from the same individual inheriting the same allele at a given locus, and a 50% chance for every generation thereafter that that particular allele will be passed on from par-

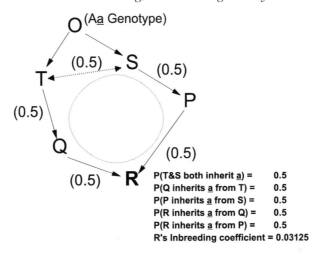

Fig. 5.1 Graphic representation of the calculation of an individual's inbreeding coefficient.

ent to offspring. Strictly speaking, the inbreeding coefficient must also take into account the level of inbreeding in the common ancestor, but pedigrees often are impossible to trace back far enough to allow precise values for inbreeding coefficients.

5.2.2 Population genetics and dominance effects

Inbreeding and heterosis tend to most directly impact traits that are associated with fitness, for reasons that are related both to population and molecular genetics. Let's examine the population genetics aspects of dominance genetic effects. To the extent that numerous production stocks can be identified and characterized, many aquaculture operations are already working with partially "inbred" animals. Inbreeding, by its very nature, is of greater concern in small populations, where random "sampling" in the formation of genotypes from an available pool of gametes can result in the eventual random exclusion of certain alleles and the fixation of others (Figs 5.2, 5.3 and 5.4).

The tendency for smaller populations to be more prone to inbreeding is exacerbated when unequal numbers of male and female broodstock are utilized. The concept of an effective population number takes gender proportions into account and reflects the relative contributions of paternal and maternal alleles to the subsequent generation (Table 5.1). The effective population number for any number of broodstock is easily calculated based on the following formula:

$$N_e = (4 N_m N_f)/(N_m + N_f),$$

where N_e is the effective population number, N_m is the actual number of breeding males, and N_f is the actual number of breeding females.

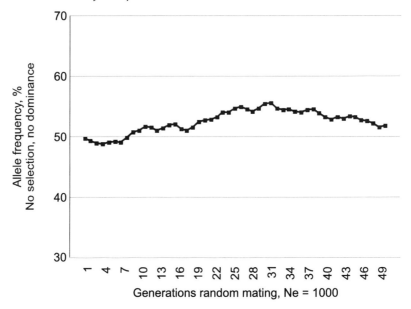

Fig. 5.2 Simulation of random drift in allelic frequency, based on an effective population size of 1000.

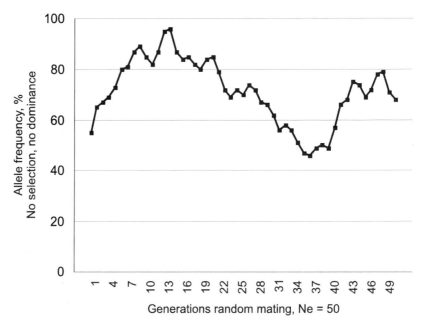

Fig. 5.3 Simulation of random drift in allelic frequency, based on an effective population size of only 50.

Inbreeding, as has previously been stated, is a relative measure. It must be gauged against some arbitrarily identified base population. The inbreeding coefficient, F, serves to describe homozygosity within

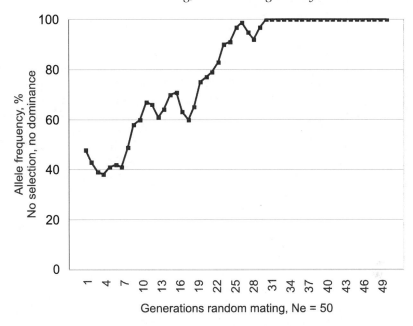

Fig. 5.4 The same simulation of random drift in allelic frequency seen in Fig. 5.3, but with the same sequence of random changes in frequency set at a slightly different starting point. Note one allele is fixed after 30 generations of random mating.

Table 5.1 Relationship of actual numbers of male and female broodstock and effective population size.

| Number broodstock | | | Effective |
Males	Females	Total	population size
5	5	10	10.00
5	10	15	13.33
5	25	30	16.67
5	50	55	18.18
10	20	30	26.67
10	40	50	32.00
50	100	150	133.33
67	67	134	134.00

a population that has resulted from inheritance of alleles that are alike by descent: alleles originating from common ancestors. If the base population for a commercial hatchery was originally 80% heterozygotic, as might be the case if two distinctly different strains were pooled, and an inbreeding coefficient of 0.30 has accumulated, then (0.30) × (80%) or 24% of the original heterozygotic loci can be assumed to have become homozygous due to inbreeding. The resultant level of heterozygosity is now 80% − 24%, or 56%. The increase in the inbreeding coefficient (F) that is expected to occur over one generation of random mating within a population of size Ne is:

$\Delta F = 1 / (2Ne)$.

Clearly, the smaller the effective population number, the greater the incremental increase in inbreeding per generation. Once inbreeding levels have accumulated within a population, they cannot be reduced simply by increasing the population number. Whenever possible, the potential for inbreeding should always be evaluated in terms of the effective, rather than actual, population number.

In traits correlated with fitness, inbreeding depression typically results from "directional dominance" at the majority of loci affecting the trait, in the direction of increasing (improving) the value. Breeding between more closely related individuals tends to increase homozygosity of all alleles when their gametes combine to form offspring. So, as small populations reproduce over a number of generations all alleles tend toward becoming fixed (Fig. 5.4). As a consequence inbreeding results in a reduction of the mean values of traits associated with fitness (specifically in the direction of the more recessive alleles) when the effects of all loci are summed. This is one way that dominance genetic effects impact values, not simply variances, of metric traits.

When discussing inbreeding and dominance effects, heterosis must be simultaneously considered. This phenomenon manifests itself when comparatively unrelated individuals produce offspring together. I use the term comparatively, because heterosis can be observed both within species (crossbreeding) and across related species (hybridization) (see 5.2.4 below). For the same reasons relating to directional dominance discussed above, traits related to fitness are usually superior in these offspring because over time certain alleles have already become fixed within parental lines or species, specifically at the myriad loci that distinguish them as distinct.

In the context of the previous statement, the term "superior" relates to a comparison with the average value of the two parental lines. In many species and many production traits, crossbred or hybrid offspring often exceed even the better-performing parental line. In relation to the concept of overdominance, the myriad loci that already distinguish the parental lines all produce multiple, rather than single, gene products. Nonetheless, crossbred and hybrid offspring are often only intermediate, and occasionally inferior, to the performance of their parental lines.

If the offspring are fertile and their effective population size can be maintained at a high enough level to avoid subsequent inbreeding depression, some portion (roughly half, theoretically) of any heterosis gained through an original cross between lines will remain in subsequent generations if random mating occurs. One downside to this approach, however, is a tremendous inflation of phenotypic variation in early generations. In spite of this tendency, this approach has been employed for centuries in breed formation of cattle, dogs and many other domesticated animals.

5.2.3 Molecular genetics and dominance effects

On a molecular level, it has also been suggested that inbreeding depression occurs because heterozygotes, which code for two gene products at a specific locus, can adapt more flexible responses to environmental variability than can homozygotes which produce only one gene product. This, of course, is an oversimplification but, in 1954, Lerner reviewed a number of studies and proposed the existence of genetic mechanisms that infer superior fitness to individuals that are heterozygous for single gene (allele) pairs or gene complexes. This phenomenon has since been referred to as "overdominance."

If this pattern of superiority (overdominance) were the rule throughout the animal kingdom it would be virtually impossible to develop highly productive inbred lines, and continuous outbreeding and crossbreeding would be the only approach available to animal breeders, including aquaculturists. The truth is, this pattern is not necessarily the rule, but it commonly comes into play to varying degrees depending on what traits, genes and species are under examination.

5.2.4 Utilizing dominance effects for genetic improvement

Within species, crossbreeding can often be utilized as a practical alternative to selection to improve traits with low heritability, and great gains can be made in some traits by this method. Nonetheless, since dominance effects are the basis for these gains, progeny must be evaluated on a cross-by-cross basis to determine how the alleles involved will interact to impact production performance for the traits in question. A practical limitation to this approach is the need to keep track of groups of broodstock (or individuals) in order to evaluate the performance of specific crosses. If one has access to enough lines within a species, and those lines can be reliably characterized, each line's potential "combining abilities" can be evaluated based on crosses with other, distinct lines.

By taking each available line and crossing it to all others, statistical analysis can be applied to estimate the general combining ability (GCA), or effect, of each line. The GCA of a line can be defined as the mean performance of that line's crossbred progeny expressed as a deviation from the mean of all crosses. Clearly, this is in large part a reflection of additive genetic effects. When two particular lines are crossed, then, the "expected" value of the progeny will be the sum of the GCAs of the parental lines. Often, however, progeny will deviate from this expected value, expressing what is referred to as the specific combining abilities (SCA) for the pair of lines being used. An SCA largely reflects dominance interactions between a number of loci involved in the expression of the trait(s) in question. Few aquaculture species, however, have been domesticated to the point where lines or strains can be characterized in terms of GCAs or SCAs.

Taking the concept of crossbreeding one step further, it is often possible to produce viable offspring by crossing closely related species.

100 Practical Genetics for Aquaculture

This practice, of course, is referred to as hybridization. While some authors refer to hybridization within a species, the term is more appropriately used for crosses between species. Unfortunately, the distinction between strains, races and species is often blurred to a point where crossbreeding or hybridization are essentially indistinguishable (Argue & Dunham 1999). While fish breeders often have specific goals in mind when they initiate hybridization studies, a significant portion of research in this area might be classified as "exploratory."

5.2.5 Alternate goals in hybridization trials

Increasingly, the expression of hybrid vigor is only one of several anticipated outcomes when species are crossed. Breeders may be looking for a cross that will produce sterile offspring to avoid unwanted reproduction in culture systems or natural habitats. In some cases, monosex populations can be produced by crossing closely related species that exhibit distinctly different sex-determining systems. Alternately, culturists may simply wish to attain some combination of desirable traits from each of the species or strains involved in a cross. In some situations, gametes from distinct species have been combined simply to see if viable offspring can be produced and characterized.

5.3 PRACTICE

5.3.1 Inbreeding impacts

Considering the behavioral hierarchies established in many aquatic species and the manifestation of these social structures under hatchery conditions, the concept of effective population size becomes an important consideration for hatchery managers throughout the world. As illustrated in Table 5.1, whenever differences arise in the relative proportion of gametes contributed by individual broodstock from generation to generation, the potential for inbreeding is increased dramatically. Considering the typical behavior of species such as tilapia, freshwater prawns, and many other aquatic organisms, it is not difficult to grasp the concept of excessive reproductive contributions by dominant male and female broodstock at the ultimate expense of genetic variation within a breeding population.

In a study of hatchery production of *Penaeus japonicus*, Sbordoni *et al.* (1986) indicated that although the number of pairs of breeding adults varied from 50 to as many as 300 per generation, genetic analyses of heterozygosity suggested effective numbers of breeding individuals may have been as low as four in some generations. Similarly, Wyban and Sweeney (1991) reported that in a study of lifetime production of nauplii among 74 female *Penaeus vannamei*, less than 10% of the females produced more than 50% of the total nauplii. In other species, particularly group spawners, similar inequities may result simply from fecundity or participatory differences. Knibb *et al.* (1998) indicated that effective population sizes in seabream (*Sparus aurata*) may at times be only 10% of actual numbers.

Inbreeding often occurs relatively unnoticed in many aquaculture settings. Common inbreeding impacts may be readily apparent, such as gross deformities, or may include more subtle effects such as reduced fecundity, survival and disease resistance. Dunham *et al*. (1987) described a number of inbreeding impacts on channel catfish (*Ictalurus punctatus*) at research facilities throughout the southeastern US. A single generation of inbreeding in this species was shown to slow egg development and decrease egg hatchability by over 15%, reduce survival to 90 days by over 5%, and reduce body weight from 30 days through 150 days by over 15%. Wada and Komaru (1994) examined impacts of inbreeding in the Japanese pearl oyster *Pinctada fucata martensii*, and noted significantly lower weight gain and survival in inbred lines selected for white shell color compared to outbred lines from crosses with brown-colored pearl oysters.

The real danger from inbreeding in most aquatic stocks, however, occurs if it is left to follow a random course. Controlled inbreeding in conjunction with selection can often be an effective approach to genetic improvement. Dunham *et al*. (1987) pointed out that fecundity in inbred channel catfish broodstock was actually higher than in randomly bred control fish in one study. At the very least, if inbreeding is unavoidable due to limited facilities it can be offset somewhat through rigorous selection for fitness-related traits. Additionally, a breeding population can be divided into two or more sub-populations, each under selection for production- and fitness-related traits. These lines can then be crossed regularly to produce fingerlings for growout purposes, in theory reducing the overall inbreeding load in the commercial stock.

5.3.2 Exploiting heterosis in a production environment

Within a species, heterosis can often be easily exploited through crossbreeding distinct, isolated strains. Dunham *et al*. (1987) listed a number of benefits attributable to crossbreeding in channel catfish. Nine of 11 crossbreeds grew better than at least one of their parent strains. Those crossbreeds that grew more rapidly than their best parental lines averaged 18% and 8% faster growth as fingerling and food-size fish, respectively. Crossbreeding and hybridization has also been shown to increase disease resistance and environmental tolerances in *I. punctatus* (Plumb *et al*. 1975; Shrestha 1977; Green *et al*. 1979; Dunham & Smitherman 1983, 1984).

Although hybridization studies often produce positive results, variable expression of dominance genetic effects often requires evaluation under a variety of conditions. In fact, the value or performance of a hybrid may be largely determined by the culture system utilized in the evaluation process. Siegwarth and Summerfelt (1992) reported that hybrid walleye (*Stizostedion vitreum* female by *S. canadense* male) were longer and heavier than pure walleyes (*S. vitreum*) when reared with overhead lighting, but not in the presence of submerged light sources. In general, submerged tank lighting reduced growth differences between walleyes and hybrids, although hybrids maintained better feed conversion and survival even in the presence of submerged lighting.

Similarly, in India, Somalingam *et al.* (1990) reported that artificially mass-produced hybrid carp fry from *Catla catla* males crossed with *Labeo rohita* females exhibited satisfactory growth rates when stocked in ponds, small reservoirs and large reservoirs, where they felt this fish might be an especially suitable candidate for artificial stocking. However, Mishra (1991) later reported that these hybrids were particularly susceptible to the parasite *Argulus siamensis* in Indian polyculture ponds.

5.3.3 Maternal effects

The choice of which sex to use from which parental line is of utmost importance in crosses where sex determination is involved, as when monosex hybrids of *Lepomis*, *Oreochromis* or *Morone* are produced. In many crossbreeding or hybridizing exercises, however, similar choices can be important for production traits as well. Commonly observed differences between reciprocal crosses of strains or species are largely attributable to "maternal effects." Accordingly, maternal effects must be accounted for when evaluating heterosis and the success of hybridization or crossbreeding activities. These effects can involve cytoplasmic inheritance from mitochondrial DNA, or more direct influences which may have genetic bases, such as egg size or quality. Inherent skill in caring for eggs or fry, as might be observed in tilapia, prawns or crayfish, also comes into play when evaluating maternal effects, and such behaviors may also be under some degree of genetic control.

Consider a practical example. One of the most widely recognized applications of hybridization in commercial aquaculture is the hybrid striped bass (Fig. 5.5), an artificially produced cross between the large, anadromous striped bass (*Morone saxatilis*) and the smaller, but hardier, freshwater white bass (*Morone chrysops*) (Fig. 5.6). Female striped bass were long preferred by hatcheries for hybridizing since they produce large numbers of eggs in a single spawning. Additionally, their eggs and larvae are larger and therefore their larvae are easier to feed and wean than those of white bass. Difficulties in collecting striped bass females and maintaining them in good spawning condition, however, have led much of the North American hybrid bass industry to rely on white bass females, which produce significantly smaller eggs but are more tolerant of capturing and handling stress and easier to collect in large numbers.

Maternal effects themselves can exhibit heterosis. Working with two strains of Nile tilapia, Tave *et al.* (1990) showed that maternal heterosis made up a substantial contribution to growth. F_2 (F_1 by F_1) and backcross (F_1 by parental line) crossbred fish were larger than their F_1 parents due to the superior maternal effects of the F_1 females resulting from the original cross. Occasionally, the influence of maternal ef-

Fig. 5.5 The hybrid striped bass (*Morone saxatilis* × *M. chrysops*, or the reciprocal) is perhaps the most widely-cultured artificially produced fish hybrid. (Photo by the author.)

Fig. 5.6 The parental species of the hybrid striped bass: *Morone chrysops* (above) and *M. saxatilis* (below). (Photo by the author.)

fects may be subtle but nonetheless quite profound in terms of behavior under culture conditions. Working with a number of tilapia hybrids, Toguyeni *et al.* (1997) described a pattern of maternal inheritance in aggressive behaviors.

5.3.4 Combining strain or species attributes

Even in the absence of significant heterosis, hybridization can often be used to combine desirable attributes of distinct species or strains. Consider an illustrative application of this approach. Basavaraju *et al.* (1995) reported on the comparative growth of reciprocal hybrids between the carps *Catla catla* and *Labeo fimbriatus*. Growth of both pure species and their reciprocal hybrids was assessed at 14-day intervals for a period of 154 days. *C. catla* attained an average final weight of 64.2 g, while *L. fimbriatus* reached only 24.1 g during the same period. Both hybrids exhibited intermediate values for final weight, suggesting a considerable influence due to additive effects, but growth of *C. catla* female by *L. fimbriatus* male offspring (with an average final weight of 57.8 g) was not significantly lower than that of the pure *C. catla*, and exceeded that of the reciprocal cross by 18 g.

Hybrids exhibited relatively small heads, characteristic of *L. fimbriatus*, as well as the deep bodies of *C. catla*, resulting in heterosis for dressout percentage. In summary, the *C. catla* female by *L. fimbriatus* male cross resulted in no significant reduction in growth, while combining desirable body conformation traits from both parental species.

5.3.5 Monosex and sterile hybrids

Hybridization can occasionally yield benefits when specific stocks can be isolated to produce all-male or all-female offspring (see Chapter 9). This is the result of closely related species having evolved different systems of sex determination. A number of sterile hybrid fish can also occasionally be found in nature. Many more can be produced artificially. Stoumboudi *et al.* (1992) reported on a naturally occurring hybrid population derived from *Barbus longiceps* and *Capoeta damascina* in Lake Kinneret in Israel. Although partially developed male- and female-like gonads were observed, all hybrids were found to be sterile. In spite of their sterile condition, hybrids exhibited a dressout percentage only intermediate to those of the parental species. Nonetheless, the authors concluded that the *Barbus* by *Capoeta* hybrid warranted evaluation as an aquaculture candidate.

The occurrence of sterile hybrids in a natural setting implies post-zygotic isolation of the parental species, rather than pre-zygotic isolation. In essence, any individual from the parental populations that produces offspring with the other species under natural conditions has wasted its gametes. These situations usually cannot continue indefinitely, and often result from deliberate or accidental translocation of non-native species or from habitat degradation and associated breakdown of physical segregation within a common watershed or reservoir.

5.3.6 Combining appropriate broodstock and gametes

Whenever hybridization or crossbreeding is to be practiced repetitively, the need to maintain populations of distinct species or lines within species is unavoidable. With this need comes the requirement for adequate facilities to house sufficient numbers of individuals within each species or line in order to avoid severe inbreeding. Care must also be taken to avoid inadvertent mixing of distinct lines within species. Just how influential the choice of strains or even individual broodstock within a species can be in hybridization studies is partially illustrated in work presented by Hulata *et al.* (1993). They evaluated no less than ten *Oreochromis niloticus* × *O. aureus* hybrids for production traits such as growth, survival, total yield and sex ratio in polyculture ponds with various carps. Obvious distinctions were observed between crosses utilizing different geographic isolates.

At times, circumstances such as recent weather patterns, handling stress and other conditions can influence the spawning success of broodstock to such an extent as to obscure genetic effects. For these and other reasons, repetition of treatments over several spawning seasons may be required to clearly determine the results of hybridization, especially when the normal spawning seasons of the species in question are somewhat out of synchrony. In some instances, distinct differences between reciprocal crosses can be determined more readily, if sufficient numbers of replicates can be utilized and myriad environmental conditions surrounding the actual production of each cross can be maintained at fairly constant levels.

Much has been learned over the years regarding the induction of spawning in various species of fish, crustaceans and mollusks. While attention to nutrition and environmental conditions may be sufficient to allow strains or even species to cross or hybridize of their own volition, special procedures are occasionally needed to promote the maturation of gametes and allow for their collection. These approaches and the associated procedures are reviewed in detail in Chapter 9.

Viability of hybrid offspring depends to a great degree on the compatibility of the karyotypes (chromosomal complements) of the parental lines involved. Reddy (1991) examined the karyomorphology (the physical characteristics of chromosomes, such as size and conformation) of grass carp, *Ctenopharyngodon idella*, silver carp, *Hypophthalmichthys molotrix*, and their hybrid. The diploid (2N) chromosome number was reported as 48 for each parental species and the hybrid, but relative proportions of metacentric, sub-metacentric and sub-telocentric pairs of chromosomes in the hybrid differed distinctly from either species. In contrast, as reported by Gorshkova *et al.* (1996), some sturgeon hybrids are the result of combining greatly different karyotypes, resulting in highly variable chromosome counts among offspring.

5.3.7 Crossbreeding or hybridization in breed formation

Often, the value of first generation hybrids is limited simply to growout for harvest. In the early 1980s, Harrell (1984) and Smith and Jenkins (1984) showed that striped bass hybrids were fertile and could produce viable offspring. Interest in spawning hybrids stemmed from the fact that traditional fingerling production based on wild-caught broodstock was so logistically challenging. The need to collect, transport, inject, hold and strip-spawn broodstock of both parental species has proven to be a major impediment to efficient fingerling production in the industry. Further pursuit of this line of inquiry by Smith et al. (1986), however, demonstrated that the F_2 progeny produced from these hybrids were not, themselves, suitable for commercial production, being much more variable phenotypically and on average only intermediate in performance when compared to their F_1 parents and the striped bass parental species. This is exactly what quantitative genetic theory would predict, an inflation of variation and simultaneous conservation of only half the heterosis derived from the original cross (Falconer 1989).

However, one interesting result from the Smith et al. (1986) study, in keeping with phenomena plant and animal breeders have utilized for centuries, was that several F_2 progeny were the fastest-growing fish in the entire study when generations were compared side by side. In all likelihood, due to the tremendous phenotypic variation in this generation, the smallest fish in the study early on were also probably F_2 progeny. Considering the cannibalistic tendencies of striped bass and their hybrids during the nursery phase, these very small F_2 fish were probably removed by their siblings before contributing to data analysis. By day 34, F_2 progeny already ranged in size from 13 to 34 mm total length.

The fact that roughly half the gains from heterosis persist indefinitely to complement any available additive genetic variation opens the door for substantial genetic improvement in populations derived from crossbred or hybrid parents. This method has been successfully utilized in production of synthetic lines of tilapia throughout the world, and in Europe and elsewhere in the development of sturgeon "breeds" derived from hybrid broodstocks (Kirpichnikov 1987; Steffens et al. 1990). In the case of hybrid striped bass, this approach would entail raising the very best performing fish from the F_2 generation (Fig. 5.7) and spawning them to produce, hopefully, a fairly uniform, high-performing production stock. Some recent work in this area has been done in Taiwan (Liu et al. 1998), but large numbers of these hybrid-derived fish might be required to offset inbreeding in such an initiative. Another problem with this approach in the genus *Morone* and other large, long-lived species involves extensive time and facility requirements (Tate & Helfrich 1998).

First generation hybrid striped bass, like many other hybrids, are comparatively uniform, being isogenic (genetically identical) for all heterozygous loci that result from combining distinct alleles previously fixed in the parental lines. The variability in subsequent generations as this condition is lost through the formation of homozygotes at these loci often allows the breeder to select for specific combinations of traits that suit his or her needs and markets. In the case of sturgeon breeding, goals may be meat production characterized by rapid growth and robust body conformation or, conversely, early maturation and associ-

Fig. 5.7 The high phenotypic variation exhibited by F_2 hybrids or crossbreds can sometimes be utilized in the development of synthetic strains or breeds. (Photo courtesy W. Lorio.)

ated caviar production, or some combination of these marketable attributes (Fig. 5.8) (Kirpichnikov 1987; Steffens et al. 1990). Researchers in Israel and the US are collaborating on similar evaluations with tilapia stocks derived from four distinct species, focusing on cold tolerance and growth.

5.4 ILLUSTRATIVE INVESTIGATIONS AND APPLICATIONS

A number of studies examining impacts of inbreeding, heterosis and hybridization in aquacultured species have been published over the years. The ultimate goal of nearly all these efforts is the improvement of performance in commercial culture, although some hybrid fish are occasionally produced for stocking into natural habitats. Growth rate

Fig. 5.8 A number of sturgeon hybrids have been produced and characterized. Some have been utilized for further development of synthetic strains. (Photo by the author.)

is probably one of the most examined traits across all species and hybrids, and, fortunately, is one of the easiest to study and analyze statistically. Other traits, however, such as disease resistance, environmental tolerances and body conformation have also been addressed in many studies.

5.4.1 A case study: carp and related species

Since the carp, *Cyprinus carpio*, is probably the single most widely cultured species on the planet, an overview of some findings related to inbreeding, crossbreeding and heterosis in this species is probably in order. In light of the low-technology conditions under which carp are often cultured and their high fecundity, most cultivated lines of this species have become somewhat inbred.

Wohlfarth (1993) provided an excellent summary of findings related to heterosis in common carp. He strictly defined heterosis as growth exceeding that of the fastest-growing parental strain, and pointed out that heterosis for growth in this species was common, but not necessarily the rule. Certain conditions, such as warmer water, and certain life-history phases such as the first-year growth period allowed for expression or quantification of heterosis in carp more readily than others. Since heterosis is quantified on a cross-by-cross and trait-by-trait basis and depends on the interaction of the parental genomes, as might be expected some crosses exhibited no heterosis. In fact, Wohlfarth indicated heterosis was entirely absent in all crosses involving one particular strain of carp.

Although heterosis is of the most value when crossbred or hybrid offspring exceed the best-performing parental line, in theory any statistically significant deviation from the mid-parent value can be considered heterosis. As noted above in the discussion of findings from Basavaraju *et al.* (1995), in some situations crossbred or hybrid offspring that perform only comparably to their best-performing parental line can still be of considerable value for culture purposes. Wiegertjes *et al.* (1995) reported on crossbreeding in common carp to evaluate heterosis for resistance to *Aeromonas salmonicida*. Survival of crossbred carp was equal to that of the better performing parental line (93%), while survival of the other parental line was considerably lower (47%). In field trials, hybrids also exhibited superior weight gain.

Working with a number of local varieties of carp (eight in all), Thien and Trong (1995) evaluated the potential for utilizing heterosis from crossbreeding to improve production in aquaculture ponds. The best crossbred carp in their trials resulted from Vietnamese white carp × introduced Hungarian scaled/mirror carp and from Hungarian scaled/mirror carp × Indonesian yellow carp. Going beyond the first generation hybrid, the authors indicated that the offspring resulting from crossing these two crossbred carp lines would be used as a synthetic population from which to develop a selection program. Their intentions were well-founded in both theory and practice. In an earlier evaluation of three common carp stocks at the Angelinka Experimental Farm in North Caucasus, Kirpichnikov *et al.* (1993) reported that a line derived eight generations earlier from a crossbreed of Ukrainian and Ropsha scaled carp exhibited higher resistance to dropsy than two comparatively inbred lines, the local mirror carp and full-scaled Ropsha carp.

Hungarian carp are often cited in works focused on crossbreeding. In a 1993 conference in Budapest, over 25 years of carp crossbreeding at the widely recognized Fish Culture Research Institute in Szarvas was summarized (Bakos & Gorda 1995). As another illustration of the hit-or-miss nature of crossbreeding, the authors pointed out that from more than 100 crossing combinations, only three particular crosses were considered to consistently produce truly outstanding results. These were classified as the Szarvas 215 mirror, the Szarvas P.31 scaly and the Szarvas P.34 scaly hybrids.

5.4.2 A case study: dominance effects in salmonids

Although a number of studies have shed light on the impact of of dominance genetic effects on performance traits in carp, more data on this topic has probably been accumulated for salmonids over the years than for any other group of fish – possibly more than for all other aquatic species combined. In an overview of breeding plans at the rainbow trout conference in Stirling in 1990, Gjedrem (1992) pointed out that although inbreeding depression is common in economically important traits of rainbow trout (*Oncorhynchus mykiss*), crossbreeding often produces variable results in this species and cannot be relied upon as a consistent method for improvement. Nonetheless, some strain crosses do produce significant improvements in this species. In keeping with the hit-or-miss nature of crossbreeding, Wangila and Dick (1996) reported on a study in which crossbreeding two distinct strains of rainbow trout resulted in 7% heterosis for growth.

Numerous hybrids between distinct salmonid species have been produced, and many have been evaluated for performance under culture conditions. Dumas *et al.* (1995) reported on the early life history of hybrids between Arctic charr (*Salvelinus alpinus*) and brook charr (*Salvelinus fontinalis*). As is sometimes the case, the choice of maternal species significantly influenced the performance of hybrid offspring. While the Arctic charr female × brook charr male fry grew at a rate comparable to the parental species during the yolk absorption period, the reciprocal hybrids grew more slowly than the other groups. Once fry began to feed, however, both hybrids grew at approximately the same rate, intermediate to the parental species. Hybrids grew faster than Arctic charr but slower than brook charr, suggesting little or no heterosis to justify this type of effort for improving production efficiencies on a commercial scale.

An interesting study evaluating the relative benefits of inbreeding and crossbreeding within a salmonid species was presented by Nihouaran *et al.* (1990) at the *Colloque sur la Truite Commune* in 1988. Evaluation of the stocking success of domestic and domestic × wild brown trout (*Salmo trutta*) in natural waters in France indicated similar success rates, although domestic trout appeared to maintain greater lengths than crossbred fish throughout the study period. Johnsson *et al.* (1994) presented another evaluation of crossing wild and domestic salmonids, conducted in Sweden. They attempted to test the hypothesis that crossing wild stocks of anadromous steelhead trout (*Oncorhynchus mykiss*) with domesticated rainbow trout (also *O. mykiss*) would reduce seasonal variation in steelhead sea-water adaptability. The authors designed a study utilizing sea-water challenges and direct transfers to sea water for 18 maternal half-sib groups of steelhead and crossbreds. Results indicated that crossing these distinct races of *O. mykiss* would indeed produce the desired results.

In a well-considered examination of genetic variation in resistance to the hemoflagellate *Cryptobia salmositica* in coho (*Oncorhynchus kisutch*) and sockeye (*O. nerka*) salmon, Bower *et al.* (1995) examined strains of each species from enzootic and non-enzootic regions as well as their intra-specific crossbreeds. When fish were challenged by inoculation, survival was highest in strains that originated in enzootic areas, lowest in strains from non-enzootic areas, and intermediate in crossbreeds. While crossbred survival was slightly above the mid-parent value in sockeye (75% vs. 70%), it fell well below the mid-parent score in cohos (30% vs. 49%). In cohos, genetic variation in resistance was not apparent within the strain that originated in the enzootic region, while additive and non-additive genetic effects were apparent in both pure strains of sockeye.

Blanc et al. (1992) reported on evaluation of a triploid hybrid between rainbow trout (*Oncorhynchus mykiss*) and Arctic charr (*Salvelinus alpinus*). By 3 years of age, hybrids did not differ significantly in weight from pure rainbow trout triploids, but were roughly 20% lighter than diploid rainbow trout. Both comparisons suggest little benefit from hybridization (or triploidization, for that matter) of these species for production purposes. Goryczko et al. (1992) reported that triploidized reciprocal crosses of *Salmo trutta* and rainbow trout generally performed comparably to diploid rainbow trout in terms of survival and growth rate. Only one cross, however, exhibited heterosis.

5.4.3 Monosex hybrid stocks

Lawrence and Morrissy (2000) hybridized a number of freshwater Australian crayfish subspecies within the yabby complex (*Cherax*). Five crosses resulted in all-male progeny, although survival was very low in two of these crosses. Three reciprocal crosses were available for comparison, but these crosses resulted in sex ratios of 1:1 or only a slight preponderance of males. Similar findings have been widely reported to occur in varied groups of fish, such as tilapia (*Oreochromis* sp.) (see Chapter 8), some North American sunfish (*Lepomis* sp.) (Childers 1967) and within lesser-known genera, such as *Rhodeus* (Duyvene de Wit 1961). Within the *Morone* complex, when female striped bass (*Morone saxatilis*) are artificially crossed with male yellow bass (*M. mississipiensis*), progeny are typically all-female.

5.4.4 Unexpected results

Contrary to the often-seen pattern of maternal influence exhibited in crosses of related teleosts, Makeyeva and Verigin (1992) reported that black carp (*Mylopharyngodon piceus*), grass carp (*Ctenopharyngodon idella*) and their hybrids ranked in the following order for growth rate during the first 3 months post-hatching in Russian ponds: pure grass carp, grass carp male by black carp female, pure black carp, and black carp male by grass carp female. In the absence of care of eggs or young, such a paternal influence on growth is somewhat more difficult to explain in practical terms than maternal influences such as egg quality or extranuclear (mitochondrial) inheritance would be.

Even stranger results sometimes arise in exploratory hybridization studies. Uthairat et al. (1993) examined the karyotypes of hybrid offspring between clariid and pangasid catfish. Offspring of crosses between *Clarias macrocephalus* and *Pangasius sutchi* were classified into two distinct intermediate morphotypes, one described as somewhat clariid-like and the other as more pangasiid-like. A third group of offspring was phenotypically indistinguishable from *C. macrocephalus*. Karyotypes for the intermediate morphotypes suggested the clariid-like offspring, with 57 chromosomes, probably possessed one set each of *C. macrocephalus* ($n=27$) and *P. sutchi* ($n=30$), while the pangasiid-like offspring, with 84 chromosomes, presumably possessed two sets from *C. macrocephalus* (2n = 54) and one set from *P. sutchi* (n = 30). The authors speculated that the offspring that were indistinguishable from *C. macrocephalus*, with 54 chromosomes, were the result of inadvertent gynogenesis (see Chapter 6).

5.4.5 Results: miscellaneous fish

Although the literature addressing inbreeding, heterosis and hybridization in aquaculture is predominated by studies focusing on carp and salmonids, many other species have also been examined. For example, Ronyai and Peteri (1990) examined phenotypic variability in growth for sterlet sturgeon (*Acipenser ruthenus*) and their hybrids (*A. ruthenus* × *A. baeri*). At 73 days, sterlet fed pellets and natural feeds weighed 15.90 ± 4.66 grams, while hybrids raised under comparable conditions weighed 26.00 ± 4.89 grams. At later life stages, hybrids continued to excel in growth in weight, by over 10% on average.

Another example of improved performance in hybrid or crossbred offspring can be seen in tilapia data presented by McGinty and Verdegem (1989). Working with Nile tilapia and Taiwanese red tilapia (Fig. 5.9), they compared growth rates of parental lines, first generation (F_1) hybrids, second generation ($F_1 \times F_1 = F_2$) hybrids, and third generation ($F_2 \times F_2 = F_3$) hybrids. Their reasons for this inquiry were straightforward enough: past work had shown that using hybrid broodstock (from a cross between female red tilapia and male Nile tilapia) resulted in improved spawning rates (as might be expected for a trait related to natural fitness).

Fig. 5.9 Virtually all strains of red tilapias are synthetic lines derived from hybrids involving *Oreochromis mossambicus*. (Photo courtesy W.R. Wolters.)

These comparisons were made both at satiation feeding levels and at approximately 85% of satiation for a 126-day growout period. The practical question to be answered was whether the growth potential of the F_2 progeny produced in this way would be sufficient for commercial production. The authors concluded that gains in fry production achieved by spawning hybrid parents were not substantially offset by any reduction in the performance of the F_2 progeny, indicating the feasibility of commercial application (Figs 5.10 and 5.11).

Tilapia stocks are often based on small founder populations, and as a result inbreeding levels can be quite high. Marengoni *et al.* (1998) described a diallel crossbreeding experiment involving three lines of Nile tilapia *Oreochromis niloticus*: Korean, Stirling and local (Japanese). After 90 days growth, overall average body weight for pure lines was approximately 70 grams, while specific heterosis for crosses ranged from 5.6 to 7.7 grams, with average heterosis of 7 grams, or roughly 10%.

Heterosis often manifests itself in general fitness and reduced susceptibility to diseases. Clayton and Price (1994) examined the influence of heterosis in poeciliid fish in terms of resistance to infection by the protozoan *Ichthyophthirius multifiliis*. The authors examined the poeciliid fish *Xiphophorus maculatus*, *X. variatus*, and their two reciprocal hybrids. After noting that overall body area and time of infection affected susceptibility, the authors adjusted the susceptibility data and determined that both reciprocal hybrids exhibited significantly lower infection levels than either parental species. Heterosis was calculated to be 16.2%, based on unadjusted values, and 31.3% after removal of body area and time of infection differences.

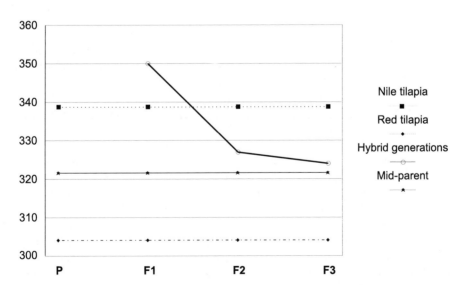

Fig. 5.10 An illustration of the comparative performance of two parental lines of tilapia and their first, second and third generation hybrids under satiation feeding. Data from a side-by-side comparison have been extrapolated to illustrate trends in performance over time for the hybrid-derived stock.

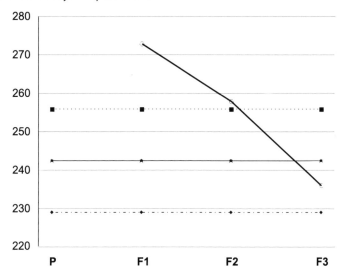

Fig. 5.11 An illustration of the comparative performance of two parental lines of tilapia and their first, second and third generation hybrids under restricted feeding. Data from a side-by-side comparison have been extrapolated to illustrate trends in performance over time for the hybrid-derived stock. Note reduced comparative performance over time under more adverse conditions.

Mukhopadhyay and Dehadrai (1987) reported differences between reciprocal *C. batrachus* hybrids. Offspring resulting from female *C. batrachus* and male vundu, *Heteropneustes fossilis*, were viable through an 8-month growout period, but the reciprocal cross exhibited severe mortality at hatching. Similarly, Hecht *et al.* (1991) related results of induced spawning of *Heteropneustes longifilis* to produce hybrids with *Clarias gariepinus*. Although purebred fry of each species exhibited comparable levels of survival, mortality was markedly increased in the *Clarias* male by *Heteropneustes* female offspring. The authors pointed out the need for further work to determine whether these findings indeed had a genetic basis.

In a straightforward example of hybrid vigor, Madu and Ita (1991) reported that African catfish hybrids resulting from a *Clarias anguillaris* female by *Clarias gariepinus* male cross outperformed either parental species in growth in both weight and length during the first 5 weeks post-hatching. Reciprocal cross fry, however, were not evaluated. Rahman *et al.* (1995) described hybridization studies in Bangladesh involving Asian catfish *Clarias batrachus* and the introduced *C. gariepinus*. When compared head-to-head for hatching rate, viability, growth and survival, *C. gariepinus* female by *C. batrachus* male offspring significantly outperformed either parental species. The reciprocal cross, however, produced mostly deformed and abnormal fry that exhibited low survival.

5.4.6 Hybrid fish for stocking natural waters

In North America, a number of stocking efforts have utilized artificially produced hybrid fish to enhance recreational fisheries opportunities in large reservoirs. Bennett *et al.*(1991) reported on the use of hybrid striped bass (*Morone saxatilis* × *M. chrysops*), tiger musky (*Esox lucius* × *E. masquinongy*), hybrid bluegill (*Lepomis* sp.) and saugeye (*Stizostedion vitreum* × *S. canadense*) to attain specific management goals for public water bodies throughout Colorado. Management objectives when stocking these fish in natural waters usually include reduction of standing biomass in undesirable species, enhanced utilization of endemic filter-feeding prey species, and increased production of trophy-sized sportfish.

Another widespread management effort in some parts of North America in recent years is the stocking of the Florida subspecies of the largemouth bass (*Micropterus salmoides floridanus*) to produce additional trophy-size fish in habitats already populated by the native (or "northern") largemouth, *M. s. salmoides*. Working in Texas, Kleinsasser *et al.* (1990) found that under identical conditions northern male by Florida female offspring were significantly heavier than parental subspecies and the reciprocal crossbred by the end of the second year of life. Pure Florida offspring were significantly smaller and in poorer condition than their crossbred or northern contemporaries. This study provides a good illustration of how dominance, maternal and additive effects can be expressed when distinct subspecies or strains within a species are forced into sympatric situations.

5.4.7 Examples: invertebrates

Lawrence and Morrissy (2000) produced a number of hybrids between freshwater crayfish subspecies within the yabby complex (*Cherax*). Half of those crosses (9 of 18) resulted in heterosis, with growth in excess of the better-performing parental line. Two crosses exhibited growth rates inferior to their slowest-growing parental lines, while seven other crosses produced intermediate growth rates. A number of crosses resulted in low juvenile production and survival and were not included in the analysis.

In a well-designed inquiry of hybridization and triploidy in oysters, Downing (1989) reported on normal and triploid progeny of *Crassostrea gigas*, *C. virginica*, and their hybrids. While one hybrid in Downing's studies was fairly easy to obtain (*C. virginica* female × *C. gigas* male), the reciprocal cross exhibited extremely low fertilization rates. The ability to utilize eggs or sperm from each individual for simultaneous production of pure or hybrid progeny allowed for a better understanding of the impact of spawning condition of individual broodstock on fertilization and viability of hybrids. Interestingly, pure line oysters generally exhibited lower mortality than did hybrids.

Rahman *et al.* (2000) presented intriguing results from hybridization studies in sea urchins of the genus *Echinometra*. In addition to additive effects reflecting species differences, performance among some hybrid crosses indicated influences of both heterosis and maternal effects.

5.5 REFERENCES

Argue, B.J. & Dunham, R.A. (1999). Hybrid fertility, introgression, and backcrossing in fish. *Reviews in Fisheries Science*, **7** (Suppl. 3–4), 137–195.

Bakos, J. & Gorda, S. (1995) Genetic improvement of common carp strains using intraspecific hybridization. In: *The Carp. Proceedings of the Second "Aquaculture"-Sponsored Symposium, Budapest, Hungary, 6–9 September 1993* (eds R. Billard and G.A.E. Gall). *Aquaculture*, **129** (Suppl. 1–4), 183–186.

Basavaraju, Y., Deveraj, K.V. & Ayyar, S.P. (1995) Comparative growth of reciprocal carp hybrids between *Catla catla* and *Labeo fimbriatus*. In: *The Carp. Proceedings of the Second "Aquaculture"-Sponsored Symposium, Budapest, Hungary, 6–9 September 1993* (eds R. Billard and G.A.E. Gall). *Aquaculture*, **129** (Suppl. 1–4), 187–191.

Bennett, C., Knox, R., Melby, J. & Stafford, J. (1991) The use of hybrid sportfish to create quality angling opportunities in Colorado. In: *Warmwater Fisheries Symposium 1, June 4–8, 1991, Scottsdale, Arizona*, pp. 99–102. USDA Forest Service, Washington, DC.

Blanc, J.-M., Poisson, H. & Vallee, F. (1992) Survival, growth and sexual maturation of the triploid hybrid between rainbow trout and Arctic charr. *Aquatic Living Resources/Ressources Vivantes Aquatiques*, **5** (Suppl. 1), 15–21.

Bower, S.M., Withler, R.E. & Riddell, R.E. (1995) Genetic variation in resistance to the hemoflagellate *Cryptobia salmositica* in coho and sockeye salmon. *Journal of Aquatic Animal Health*, **7** (Suppl. 3), 185–194.

Childers, W.F. (1967) Hybridization of four species of sunfishes (Centrarchidae). *Bulletin of the Illinois Natural History Survey*, **29** (Suppl. 3), 158–214.

Clayton, G.M. & Price, D.J. (1994) Heterosis in resistance to *Ichthyophthirius multifiliis* infections in poeciliid fish. *Journal of Fish Biology*, **44** (Suppl. 1), 59–66.

Downing, S.L. (1989) Hybridization, triploidy and salinity effects on crosses with *Crassostrea gigas* and *Crassostrea virginica*. *Journal of Shellfish Research*, **8** (Suppl. 2), 447.

Dumas, S., Blanc, J.M., Audet, C. & De la Nouee, J. (1995) Variation in yolk absorption and early growth of brook charr *Salvelinus fontinalis* (Mitchill), Arctic charr, *Salvelinus alpinus* (L.), and their hybrids. *Aquaculture Research*, **26** (Suppl. 10), 759–764.

Dunham, R.A. & Smitherman, R.O. (1983) Relative tolerance of channel × blue hybrid and channel catfish to low oxygen concentrations. *The Progressive Fish-Culturist*, **45** (Suppl. 1), 55–56.

Dunham, R.A. & Smitherman, R.O. (1984) Ancestry and breeding of catfish in the United States. *Circular 273*, Alabama Agricultural Experiment Station. Auburn University.

Dunham, R.A., Greenland, D., Kincaid, H.L., Reagan, R.E., Smitherman, R.O. & Wolters, W.R. (1987) Genetics and breeding of catfish. *Southern Cooperative Series Bulletin 325*, Auburn University.

Duyvene de Wit, J.J. (1961) Hybridization experiments in rhodeine fishes (Cyprinidae, Teleostei). An intergeneric hybrid between female *Rhodeus ocellatus* and male *Acanthorodeus artemius*. *Zoologica* (New York), **46,** 25–26.

Falconer, D.S. (1989) *Introduction to Quantitative Genetics*, 3rd edn. Longman, Essex.

Gjedrem, T. (1992) Breeding plans for rainbow trout. In: *The rainbow trout. Aquaculture*, **100** (Suppl. 1–3), 73–83.

Gorshkova, G., Gorshkov, S., Gordin, H. & Knibb, W. (1996) Karyological studies in hybrids of beluga *Huso huso* and the Russian sturgeon *Acipenser gueldenstaedti* Brandt. *Israeli Journal of Aquaculture/Bamidgeh*, **48** (Suppl. 1), 35–39.

Goryczko, K., Dobosz, S., Luczynski, M. & Jankun, M. (1992) Triploidized trout hybrids in Aquaculture. *Bulletin of the Sea Fisheries Institute, Gdynia*, **125**, 3–6.

Green, O.L., Smitherman, R.O. & Pardue, G.B. (1979) Comparisons of growth and survival of channel catfish, *Ictalurus punctatus* from distinct populations. In: *Advances in Aquaculture* (eds T.V.R. Pillay & W.A. Dill), pp. 626–628. Fishing News Books, Ltd., Farnham, Surrey.

Harrell, R.M. (1984) Tank spawning of first generation striped bass × white bass hybrids. *The Progressive Fish-Culturist*, **46**, 75–78.

Hecht, T., Lublinkhof, W. & Kenmuir, D. (1991) Induced spawning of the vundu *Heteropneustes longifilis*, and embryo survival rates of pure and reciprocal clariid crosses. *South African Journal of Wildlife Resources*, **21** (Suppl. 4), 123–125.

Hulata, G., Wohlfarth, G.W., Karplus, I. et al. (1993) Evaluation of *Oreochromis niloticus* × *O. aureus* hybrid progeny of different geographical isolates, reared under varying management regimes. *Aquaculture*, **115** (Suppl. 3–4), 253–271.

Johnsson, J.I., Clarke, W.C. & Blackburn, J. (1994) Hybridization with domesticated rainbow trout reduces seasonal variation in seawater adaptability of steelhead trout (*Oncorhynchus mykiss*). In: *Salmonid Smoltification 4. Aquaculture*, **121** (Suppl. 1–3), 73–77.

Kirpichnikov, V.S. (1987) Fish culture and new breeds of pond fish in the USSR. *Journal of Ichthyology*, **27** (Suppl. 3), 89–98.

Kirpichnikov, V.S., Ilyasov, I, Jr, Shart L.A. et al. (1993) Selection of krasnodar common carp (*Cyprinus carpio* L.) for resistance to dropsy: Principal results and prospects. In: *Genetics in Aquaculture IV* (eds G.A.E. Gall & H. Chen). *Aquaculture*, **111**, 7–20.

Kleinsasser, L.J., Williamson, J.H.& Whiteside, B.G. (1990) Growth and catchability of northern, Florida and F_1 hybrid largemouth bass in Texas ponds. *North American Journal of Fisheries Management*, **10** (Suppl. 4), 462–468.

Knibb, W.R., Gorshkov, S. & Gorshkova, G. (1998) Genetic improvement of cultured marine fish: case studies. In: *Tropical Mariculture* (ed. S. De Silva), pp. 111–149. Academic Press, Sydney.

Lawrence, C.S. & Morrissy, N.M. (2000). Genetic improvement of marron *Cherax tenuimanus* Smith and yabbies *Cherax* spp. in Western Australia. *Aquaculture Research*, **31** (Suppl. 1), 69–82.

Lerner, I.M. (1954) *Genetic Homeostasis*. John Wiley and Sons, Inc., New York.

Liu, F.-G., Cheng, S.-C. & Chen, H.-C. (1998) Induced spawning and larval rearing of domestic hybrid striped bass (*Morone saxatilis* × *M. chrysops*) in Taiwan. *Israeli Journal of Aquaculture/Bamidgeh*, **50** (Suppl. 3), 111–127.

Madu, C.T. & Ita, E.O. (1991) Comparative growth and survival of hatchlings of *Clarias* sp., *Clarias* hybrid and *Heterobranchus* sp. in the indoor hatchery. *Annual Report of National Freshwater Fisheries Research* (Nigeria), **1990**, 47–50.

Makeyeva, A.P. & Verigin, B.V. (1992) Morphological characters of age 0+ reciprocal hybrids of grass carp, *Ctenopharyngodon idella*, and black carp, *Mylopharyngodon piceus*. *Journal of Ichthyology* (Russia), **33** (Suppl. 3), 66–75; *Voprosy Ikhtiologii*, **32** (Suppl. 6), 41–48.

Marengoni, N.G., Onoue, Y. & Oyama, T. (1998) Offspring growth in a diallel crossbreeding with three strains of Nile tilapia *Oreochromis niloticus*. *Journal of the World Aquaculture Society*, **29** (Suppl. 1), 114–119.

McGinty, A.S. & Verdegem, M.C. (1989) Growth of *Tilapia nilotica*, Taiwanese red tilapia, and their F1, F2 and F3 hybrids. *Journal of the World Aquaculture Society*, **20** (Suppl. 1), 56A.

Mishra, B.K. (1991) Observations on the susceptibility of the hybrid, *Labeo rohita* × *Catla catla* to *Argulus siamensis* Wilson infection in composite culture ponds. In: *Proceedings of the National Symposium on New Horizons in Freshwater Aquaculture*, pp. 184–185. Central Institute of Freshwater Aquaculture, Kausalyagang, India.

Mukhopadhyay, S.M. & Dehadrai, P.V. (1987) Survival of hybrids between air-breathing catfishes *Heteropneustes fossilis* (Bloch) and *Clarias batrachus* (Linn.). *Matsya*, **12–13**, 162–164.

Nihouaran, A., Porcher, J.P., Maisse, G. & Bagliniere, J.L. (1990) Comparaison des performances de deux souches de truite commune, *Salmo trutta* L., (domestique et hybride sauvage × domestique) introduites au stade alevin dans un rulsseau. In: *Actes du Colloque sur la Tuite Commune, Centre du Paraclet, Sept. 1988, 2e Partie* (eds G. Maisse and E. Vigneux). *Bulletin Francaise de la Peche et de la Pisciculture*, **319-SP**, 173–180.

Plumb, J.A., Green, O.L., Smitherman, R.O.& Pardue, G.B. (1975) Channel catfish virus experiments with different strains of channel catfish. *Transactions of the American Fisheries Society*, **104**, 140–143.

Rahman, M.A., Bhadra, A., Begum, N., Islam, M.S. & Hussain, M.G. (1995) Production of hybrid vigor through cross breeding between *Clarias batrachus* Lin. and *Clarias gariepinus* Bur. *Aquaculture*, **138** (Suppl. 1–4), 125–130.

Rahman, M.A., Uehara, T. & Aslan, L.M. (2000) Comparative viability and growth of hybrids between two sympatric species of sea urchins (Genus *Echinometra*) in Okinawa. *Aquaculture*, **183** (Suppl. 1–2), 45–56.

Reddy, P.V.G.K. (1991) A comparative study of the karyomorphology of grass carp, *Ctenopharyngodon idella* (Val.), silver carp *Hypophthalmichthys molitrix* (Val.) and their F_1 hybrid. *Bhubaneswar*, **1**, 31–41.

Ronyai, A. & Peteri, A. (1990) Comparison of growth rate of sterlet (*Acipenser ruthenus* L.) and hybrid of sterlet × Lena River's sturgeon (*A. ruthenus* L. × *A. baeri stenorhynchus* Nikolsky) raised in a water recycling system. *Aquaculture Hungary*, **6**, 185–192.

Sbordoni, V., De Matthaeis, E., Cobolli Sbordoni, M., La Rosa, G. & Mattoccia, M. (1986) Bottleneck effects and the depression of genetic variability in hatchery stocks of *Penaeus japonicus* (Crustacea, Decapoda). *Aquaculture*, **57**, 239–251.

Shrestha, S.B. (1977) *The parasites of different strains and species of catfish (Ictalurus sp.)*. MS Thesis, Auburn University.

Siegwarth, G.L. & Summerfelt, R.C. (1992) Light and temperature effects on performance of walleye and hybrid walleye fingerlings reared intensively. *The Progressive Fish-Culturist*, **54** (Suppl. 1), 49–53.

Smith, T.I.J. & Jenkins, W.E. (1984) Controlled spawning of F1 hybrid striped bass (*Morone saxatilis*) × (*M. chrysops*) and rearing of F2 progeny. *Journal of the World Mariculture Society*, **14**, 147–161.

Smith, T.I.J., Jenkins, W.E. & Snevel, J.F. (1986) Production characteristics of striped bass (*Morone saxatilis*) and F1, F2 hybrids (*M. saxatilis* and *M. chrysops*) reared in intensive tank systems. *Journal of the World Mariculture Society*, **16**, 57–70.

Somalingam, J., Maheshwari, U.K. & Langer, R.K. (1990) Mass production of intergeneric hybrid catla (*Catla catla* male × *Labeo rohita* female) and its growth in ponds and small and large reservoirs of Madya Pradesh. In: *Carp Seed Production Technology* (eds P. Keshavanath & K.V. Radhakrishnan), pp. 49–52. Asian Fisheries Society, Indian Branch, No. 2.

Steffens, W., Jachinen, H. & Fredrich, F. (1990) Possibilities of sturgeon culture in central Europe. *Aquaculture,* **89,** 101–122.

Stoumboudi, M., Villwock, W., Golenser, E., Abraham, M. & Moav, B. (1992) The *Barbus longiceps/Capoeta damascina* hybrids – qualified for aquaculture? In: *Progress in Aquaculture Research,* pp. 197–204. Special Publication no 17, European Aquaculture Society.

Tate, A.E. & Helfrich, L.A. (1998) Off-season spawning of sunshine bass (*Morone chrysops* × *M. saxatilis*) exposed to 6- or 9-month phase-shifted photothermal cycles. *Aquaculture,* **167** (Suppl. 1–2), 67–83.

Tave, D., Smitherman, R.O., Jayaprakas, V. & Kuhlers, D.L. (1990) Estimates of additive genetic effects, maternal genetic effects, individual heterosis, maternal heterosis, and egg cytoplasmic effects for growth in *Tilapia nilotica*. *Journal of the World Aquaculture Society,* **21** (Suppl. 4), 263–270.

Thien, T.M. & Trong, T.D. (1995) Genetic resources of common carp in Vietnam. In *The Carp. Proceedings of the Second "Aquaculture"-Sponsored Symposium, Budapest, Hungary, 6–9 September 1993* (eds R. Billard & G.A.E. Gall). *Aquaculture,* **129** (Suppl. 1–4), 216.

Toguyeni, A., Fauconneau, B., Melard, C. *et al.* (1997) Sexual dimorphism studies in tilapias, using two pure species, *Oreochromis niloticus* and *Sarotherodon melanotheron,* and their intergeneric hybrids (*O. niloticus* × *S. melanotheron* and *S. melanotheron* × *O. niloticus*). In: *Tilapia Aquaculture: Proceedings from the Fourth International Symposium on Tilapia in Aquaculture* (ed. K. Fitzsimmons), pp. 200–212. NRAES-106 Volume 1, Northeast Regional Agricultural Engineering Service, Ithaca, New York.

Uthairat, N.-N., Prasit, T.W. & Roberts, T.R. (1993) Chromosome study of hybrid and gynogenetic offspring of artificial crosses between members of the catfish families Clariidae and Pangasiidae. *Environmental Biology of Fishes,* **37** (Suppl. 3), 317–322.

Wada, K.T. & Komaru, A. (1994) Effect of selection for shell coloration on growth rate and mortality in the Japanese pearl oyster, *Pinctada fucata martensii*. *Aquaculture,* **125** (Suppl. 1–2), 59–65.

Wangila, B.C.C. & Dick, T.A. (1996) Genetic effects and growth performance in pure and hybrid strains of rainbow trout, *Oncorhynchus mykiss* (Walbaum) (order: Salmoniformes, family: Salmonidae). *Aquaculture Research,* **27** (Suppl. 1), 35–41.

Wiegertjes, G.F., Groeneveld, A. & Van Muiswinkel, W.B. (1995) Genetic variation in susceptibility to *Trypanoplasma borreli* infection in common carp (*Cyprinus carpio* L.). *Veterinary Immunology and Immunopathology,* **47** (Suppl. 1–2), 153–161.

Wohlfarth, G.W. (1993) Heterosis for growth rate in common carp. *Aquaculture,* **113** (Suppl. 1–2), 31–46.

Wyban, J.A. & Sweeney, J.N. (1991) *Intensive Shrimp Production Technology. The Oceanic Institute Shrimp Manual*. The Oceanic Institute, Makapuu Point, Honolulu, Hawaii.

Chapter 6
Chromosomal Genetics I: Gynogenesis and Androgenesis

6.1 INTRODUCTION

Although the title of this chapter refers to chromosomal manipulation, the key concept in this discussion is the manipulation of entire sets, or complements, of chromosomes rather than individual strands of DNA. Gynogenesis and androgenesis are based on limiting inheritance to the maternal or paternal parent, respectively, most notably by altering normal processes in the creation and replication of diploid (2N) genomes in developing zygotes. The key to both these approaches lies in eliminating the genetic contribution of one parent or the other and subsequently interrupting the normal course of events in a newly-fertilized or activated egg in such a way (or ways) as to result in the restoration of a 2N state, allowing normal development to proceed.

Although such approaches are not feasible with higher vertebrates, viable offspring have been produced through gynogenesis and androgenesis in many fish and invertebrates. Gynogenesis and androgenesis have little direct application in the production of commercial stocks, but they can be utilized to ascertain mechanisms of sex determination as well as to produce monosex populations, large numbers of genetically identical clones, and novel populations with far greater levels of homozygosity than might ever be found in nature.

The fact that many aquatic species can survive gynogenesis has also proven a valuable tool for studies of gene linkage, sources of phenotypic variation and overall improvement of production traits through directed breeding. An understanding of gynogenesis and androgenesis should lay the groundwork for discussion of polyploidy, another interesting area of aquaculture genetic research, in Chapter 7.

6.2 THEORY

6.2.1 Meiosis and polar bodies

The vast majority of the fish, crustaceans and bivalves aquaculturists concern themselves with possess diploid (2N) genomes, consisting of two homologous "sets" (each of which is denoted as 1N, or "haploid"). Chromosomes typically occur in pairs within cell nuclei; the number of pairs is the same as the number of distinct chromosomes in each 1N set (see Chapter 2). The number and physical characteristics of chromosomes are typically defining characteristics of a species. In humans, a normal complement (2N) is comprised of 23 pairs of chromosomes; in goldfish, 47 pairs; in alligators, 16 pairs. Normally, one of the homologous chromosome sets is of maternal origin, and the other is of paternal origin. As the organism in question reaches maturity and produces its own gametes, the two (1N) sets of chromosomes must be separated again to arrive at one (1N) set in each gamete. While Chapter 2 provided a brief review of the mechanisms by which 1N gametes are formed within 2N organisms, several processes must now be addressed in more detail. In keeping with the practical focus of this text, however, the discussion will be somewhat oversimplified.

During the formation of gametes, each pair of homologous chromosomes, or homologues, is separated. Within a pair, the chromosome of paternal origin and that of maternal origin are destined to occupy different gametes. Each pair, however, separates independently and ultimately, two copies of the genetic material on each original chromosome find their way into gametes (or polar bodies, as will be covered below). Additionally, during the process of meiosis material from one strand of DNA can be exchanged with the corresponding region of a homologous strand. These processes collectively result in (1N) gametes with myriad combinations of genetic material from the original paternal and maternal (1N) chromosome sets.

In sperm production, the original cell contents in the primary spermatocyte are typically divided more-or-less evenly between the secondary spermatocytes and subsequently among the four resultant sperm cells. In the case of egg formation, however, cell contents are more-or-less conserved throughout the process of oogenesis and are utilized by only one of the four haploid sets of chromosomes. While this solitary (1N) genome ultimately persists to combine with a sperm cell in the formation of a new organism, the other three chromosome sets are lost, in what are referred to as the first and second polar bodies. This certainly makes sense from an evolutionary standpoint, albeit within the framework of species life history and fecundity, since in most aquatic species a larger egg results in a more viable offspring.

The first polar body represents the first two "excess" (1N) genomes discarded as the 4N primary oocyte transforms into a 2N secondary oocyte. The first polar body is generally lost early in the process of egg development in most fish, but in mollusks it typically remains within the developing oocyte. The second polar body includes the third "excess" haploid genome. It is generally retained within the oocyte until fertilization takes place. At this point, mechanisms triggered by the penetration of sperm result in "activation" of the oocyte and the expulsion of the second polar body and its (1N) chromosome set, leaving the final (1N) chromosome set to combine with the (1N) genetic material introduced via the sperm (Fig. 6.1).

Expulsion of the second polar body can be considered the final step of meiosis. The retention of the second polar body and the genetic material it contains until after the egg has been fertilized is the process that allows for production of one type of gynogenetic offspring. As we will discuss further, in "meiotic," or "heterozygous," gynogenesis this final step of meiosis is prevented from occurring. However, in order to utilize the genetic material derived only from the maternal parent, the first requirement is to treat, manipulate or substitute sperm in such a way as to prevent any paternal genetic contribution.

6.2.2 Meiotic gynogenesis

Early studies of gynogenesis relied on high levels of radiation to destroy the genetic material within sperm while still keeping them sufficiently viable to allow for motility and egg penetration. Researchers later found they could achieve the same goals more safely and easily with ultraviolet (UV) light. Milt is typically spread in a thin layer in a glass dish or plate and kept from overheating (often by a cool water bath surrounding the dish or plate) while the appropriate exposure level is applied. Exposure is often determined by first estimating a level at which viability is completely eliminated. Common UV germicidal sterilizing lamps can be used, and UV intensity can be measured to calculate the dose using a standard UV meter. Additionally, sperm from entirely different species can sometimes be utilized to further prevent complications from incorporation of intact bits of paternal genetic material.

Ovulated eggs are then exposed to this biologically active but genetically "blank" milt. As a sperm cell activates an egg, intracellular mechanisms to eject the second polar body are set into motion. At this

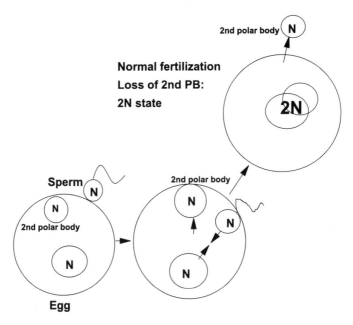

Fig. 6.1 The normal sequence of events during fertilization in most aquatic species.

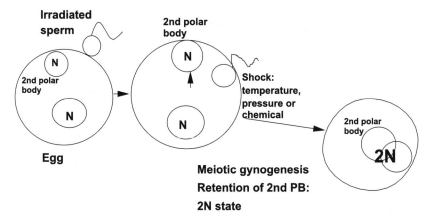

Fig. 6.2 Mechanisms of meiotic gynogenesis.

critical point in time a physiological shock such as heat, cold or pressure must be applied to halt the loss of the second polar body, the last step of meiosis. This allows the egg to proceed with normal development utilizing the two sets of chromosomes it subsequently possesses (Fig. 6.2). This process is gynogenesis (gyno- genesis), and offspring produced in this way are typically referred to as "gynogens."

6.2.3 Mitotic gynogenesis

While the diploid meiotic gynogens produced in the manner outlined above contain genetic material only from their maternal parents, they cannot be considered clones. Although each chromosome set is of maternal origin, they are not identical. Clones can, however, be produced through gynogenesis but they cannot be considered clones of the original maternal parent. The basis for this process is the second method of gynogenesis: mitotic gynogenesis. Consider a variation of the meiotic gynogenesis process described above, where irradiated sperm is used to activate egg cells but polar bodies are allowed to be jettisoned. The result is a collection of activated eggs, each with a unique 1N set of chromosomes provided by the maternal parent. Normally, virtually all of these haploid eggs would begin development through the usual process of chromosome replication and cell division (mitosis), only to die at some point early in the process as a result of gross deformities and general lack of vigor.

By halting the normal mitotic processes involved in the first cell division of these haploid, activated eggs at the precise time, they can be transformed (albeit in occasionally very low numbers) into viable 2N organisms. The key to this approach is to apply a physiological shock (again, through application of heat, cold or pressure) during the window of time between the replication of the 1N chromosomal complement and the subsequent separation of the chromosomes and cellular contents into two distinct cells (Fig. 6.3). When successful, this process

124   *Practical Genetics for Aquaculture*

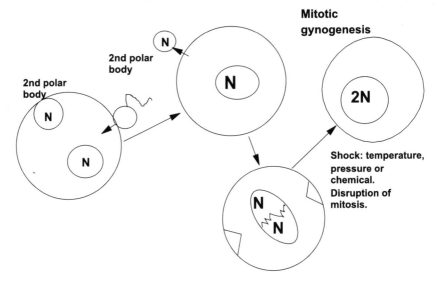

Fig. 6.3   Mechanisms of mitotic gynogenesis.

results in: (1) disruption of the mechanical division of the cell, (2) restoration of a 2N diploid condition within the undivided cell, and perhaps most interestingly, (3) homozygosity at every locus in the entire genotype.

Accordingly, mitotic gynogenesis is also sometimes referred to as "homozygous" gynogenesis. If these homozygous 2N fish are females (from a species with female homogameity and no complicating genetic or environmental influences), their eggs, when mature, could then be subjected to gynogenesis. While many genetically identical offspring could then be produced from a single organism, each line of clones would be unique to the individual female gynogen utilized to produce them. In theory (and often in practice), some of these daughter clones can be masculinized, through exposure to or ingestion of hormones early in their development, and utilized for production of an even larger subsequent generation of clones.

6.2.4   Androgenesis

The rationale and mechanisms associated with androgenesis are the same as those in mitotic gynogenesis. The sole difference, in practice, is the destruction of genetic material within the mature oocyte, rather than the sperm, and the subsequent interruption of mitosis to allow for a 2N homozygous organism with genetic material derived solely from the sperm's contribution (Fig. 6.4). One potential problem with this approach is the destruction not only of the nuclear genetic material but also of structures associated with RNA and amino acid synthesis, as well as mitochondrial DNA.

One additional approach to both gynogenesis and androgenesis that should be mentioned at this point is the use of (2N) gametes

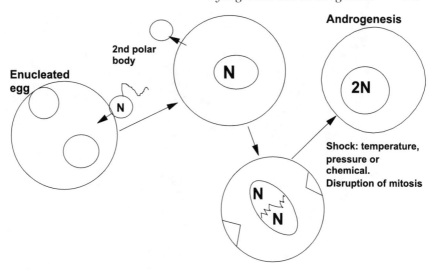

Fig. 6.4 Typical method for production of androgenetic aquatic organisms.

from tetraploid (4N) organisms. In theory, genetically inert oocytes or sperm, produced through irradiation, can be united with 2N gametes to result in normally developing organisms. In this way, a "heterozygous" form of androgenesis is possible. We will discuss the production and potential of tetraploid organisms in Chapter 7.

6.3 PRACTICE

Although work relating to androgenesis in aquatic species is proceeding in many parts of the world, many more practical applications of gynogenesis are found in the literature. Some examples of naturally occurring gynogenesis can be found in teleost fish, but most gynogenesis of concern to aquaculturists occurs under laboratory conditions (Fig. 6.5). The reasons for inducing gynogenesis range from production of monosex populations to development of partially or completely inbred organisms. Partially inbred offspring (from meiotic gynogenesis) may be useful in genome studies examining the position of various loci in relation to the centromeres of their chromosomes. When true clones can be produced in large numbers through gynogenesis these genetically identical offspring can be of great value in studies relating to nutrition, disease resistance or other topics.

6.3.1 Heterozygous versus homozygous gynogenesis

The meiotic:heterozygous and mitotic:homozygous gynogenetic associations deserve more discussion before reviewing recent work in these areas. Recall that heterozygous refers to the presence of two distinct forms (alleles) of a gene, while homozygous implies two copies of

Fig. 6.5 Ornamental koi carp (*Cyprinus carpio*) have been used in gynogenetic studies for determination of colors and coloration pattern inheritance, as well as in androgenetic studies as a source of easily-detected coloration to identify parental genetic contribution(s). (Photo by the author.)

the same form of the gene. Albinism, for example, is a good illustration of a homozygous condition, since albinos typically have two copies of a recessive allele at one particular locus that "short circuits" the normal pathways for pigment development (Chapter 2). If only one copy is inherited, the animal will express normal pigmentation and merely be a carrier, passing along the albinism gene to roughly half of its progeny. Albino fish and their progeny lend themselves particularly well to the study of gynogenesis, for reasons that will become apparent (see 6.3.3 below).

In meiosis, chromosomes inherited from either of an organism's parents (one generation back) can mingle and end up together in a gamete (one generation forward). That is, if an organism normally has a total of eight chromosomes, a set of four from each of its parents, chromosomes 1 and 2 from the organism's mother could end up in the same gamete with chromosomes 3 and 4 from the organism's father. Any other combination could also arise, so long as each gamete has one of each of the four chromosomes that make up a complete set. Each chromosome in a complement carries specific codes for certain gene products and has its own distinct characteristics in terms of size and shape. Chromosomes within a complement are not interchangeable, although unfortunately they can seem to appear that way when scientists try to enumerate and characterize them under the microscope. In this simple hypothetical situation ($1N = 4$), when meiotic gynogenesis takes place, the odds of the second polar body and the oocyte nucleus having the same origins (maternal or paternal from the previous generation) for all of the chromosomes they each contain is slim ($0.5 \times 0.5 \times 0.5 \times 0.5 = 1$ in 16).

Consider a single heterozygous locus in the maternal parent: egg pronuclei could include either allele, as could second polar bodies, and resultant meiotic gynogens could potentially be homozygous for either allele or heterozygous. In those chromosome pairs with different origins, at least some portion of the alleles will also be different. So in most instances we might encounter, even without the phenomena of crossing over and sister chromatid exchange (which we need not go into for this discussion), an organism produced through meiotic gynogenesis will be heterozygous for some measurable portion of its genome. Hence, the term "heterozygous" gynogenesis.

6.3.2 Determining appropriate procedures

At this point, an examination of the mechanics of gynogenesis in aquacultured species is required. Protocols previously illustrated for carrying out gynogenesis involve (1) activating eggs with sterilized sperm or sperm from unrelated species, and either (2) applying a physiological shock (heat, cold or pressure) or a chemical (such as cytochalasin B in bivalves) to prevent expulsion of the second polar body, or (3) waiting until the second polar body has been expelled and the single remaining chromosome complement has completed its initial replication, then applying a physiological shock to prevent the first cell division. While these techniques sound fairly straightforward, in practice gynogenesis is very hit-or-miss in nature and anything but a cookbook exercise.

Much of the work to date in this area has focused on defining the parameters of sperm irradiation as well as the intensity, duration and timing of physiological shocks for various species. For example, working in India with the ornamental labyrinth fish *Betta splendens*, Kavumpurath and Pandian (1994) examined the use of irradiated tilapia sperm and hydrostatic pressure shocks to produce gynogens. Survival of heterozygous gynogens was maximized, at 50%, when 7000 psi was applied for 6 min duration, 2.5 min after activation. In contrast, homozygous gynogens were produced at a rate of 21% in the same study when a 5-min pressure shock was applied at 34 min post-activation. The results imply that under these specific conditions, the second polar bodies of these *Betta splendens* eggs were typically ejected around 2.5–8.5 min post-activation, and that by roughly 26 min later, the first replication of chromosomes had taken place but the original cell had not yet divided. Similarly, in work with a Japanese loach, *Misgurnus anguillicaudatus*, Suwa et al. (1994) reported successful meiotic gynogenesis based on pressure shock at 5 min post-activation and mitotic gynogenesis resulting from pressure shock at 30–35 min post-activation.

Timing of shock treatments is, of course, of equal importance in androgenesis, and two examples from the literature serve to further illustrate the relationship of timing of these events to biological processes and prevailing temperatures. Working with the loach, *Misgurnus anguillicaudatus,* Masaoka *et al.* (1995) used a pressure shock (800 kg cm^{-2}) for 1 min to interrupt mitosis 35 min after activation with viable sperm. Eggs had been denucleated via UV (7500 erg mm^{-2}). In contrast, in the cold-water salmonid *Oncorhynchus masou,* Nagoya *et al.* (1996) used a pressure shock (650 kg cm^{-2}) for 6 min to interrupt mitosis, 450 min after fertilization of eggs that had been exposed to Cobalt-60 (450 Gy). Since the development and evaluation of induction protocols requires a knowledge of the precise time of egg activation or fertilization, gametes used in gynogenesis or androgenesis are usually taken and combined artificially (see Chapter 9).

The most common technique for applying hydrostatic pressure shocks to eggs of aquatic organisms involves the use of a stainless steel cylinder which is bored out to create a one-piece vessel with an open well extending for most of its length. A brass piston with an O-ring is closely fitted to match the internal diameter of the bored cylinder, and a pressure gauge and relief valve are incorporated into either the piston or, more often, the cylinder. Eggs are placed into the cylinder, and the piston is partially inserted until a seal is achieved. The entire device is then placed into a hydraulic press, which is used to apply pressure to drive the piston into the cylinder. Pressure is raised rapidly to the desired level and held for a specified period of time, at which point it can be released instantaneously. Clearly, this practice can pose a certain degree of risk, so appropriate measures and precautions should be taken to prevent injury to personnel in the event of an accident or cylinder failure.

Cold-shocks are the normal manipulation approach for disrupting meiosis or mitosis in warm-water species; for cool- or cold-water species, heat shocks are commonly applied. Taniguchi *et al.* (1986) described protocols used to successfully induce meiotic gynogenesis in common carp (*Cyprinus carpio*). After activating ova with milt previously exposed to 9000+ erg mm^{-2}, they used a 60-min cold shock of 0–0.5°C applied 10–12 min post-activation (at an incubation temperature of 20°C) to prevent the expulsion of the second polar body. Conversely, in the cold-water salmonid *Oncorhynchus mykiss,* Diter *et al.* (1993) found heat shock to be an effective approach to the production of mitotic gynogens. From a wide range of possibilities requiring evaluation (27–33° thermal shock, 2–30 min shock duration and 2–4 h post-activation shock initiation), applications of 30°C for 9 min, 31°C for 5 min, or 32°C for 4 min applied between 3 and 4 h post-activation gave the best results.

The physiological shocks used in androgenesis and gynogenesis are often difficult to standardize with precision. Baths for heat- or cold-shock should be thermostatically controlled and circulated by pumps or aeration to provide uniform temperature differentials. Rates of pressure increase or release may vary depending on equipment and procedures utilized, and temperature shock effects may not only reflect the thermal conductance characteristics of containers used to immerse eggs (netting, plastic bowls, polyethylene bags, etc.) but also various water chemistry variables (hardness, alkalinity, pH, etc.).

Tau$_0$ and bimodal optima

Since water temperature is one of the major variables effecting the timing of the activation → second polar body ejection → initial chromosome replication → first cleavage course of events, some researchers have tried to describe results of gynogenetic studies in terms of a dimensionless parameter called tau$_0$, defined as the duration of one mitotic cycle of cell division as related to water temperature. For example, in the tench *Tinca tinca*, Linhart *et al.* (1995) reported that at an incubation temperature of 20°C, physiological shocks at 2, 5, and 10 min post-activation corresponded to 0.069, 0.173 and 0.346 tau$_0$, respectively.

Cherfas *et al.* (1992) reported on experiments to determine the optimal timing of meiotic and mitotic gynogenesis in koi carp (*Cyprinus carpio*). They found the best results were obtained from cold shock at 1–2 min post-activation (tau$_0$ = 0.03–0.07) and from heat shock at 27–29 min post-activation (tau$_0$ = 1.5–1.6). In a subsequent study, Cherfas *et al.* (1994) described distinct differences in the gynogenetic success and timing of different physiological shocks. Working again with koi carp, they noted that cold shocking produced the most viable meiotic gynogens at 0.05–0.10 tau$_0$, along the lines of their previous findings, but also again at 0.30–0.40 tau$_0$. Poor induction resulted from cold shocks in the period separating the two peaks. In contrast, heat shock exhibited a single meiotic gynogenesis optimum of 0.15–0.25 tau$_0$.

Refining these numbers further in a follow-up report, Cherfas *et al.* (1995b) defined tau$_0$ = 1.0 for common carp as 29.4 min at 20°C and as 20.4 min at 24°C. However, Sumantadinata *et al.* (1990a) found previously that optimal applications of heat shock to produce meiotic and mitotic gynogens in the Indonesian strain of common carp at 25°C occurred at 2–4 min and 40–45 or 55 min, respectively, suggesting a variance in the tau$_0$ scale from that established in Israeli ornamental carp. This could be explained by the assumption that tau$_0$ must be established empirically for the species and conditions unique to each study, further complicating the practicality of this approach.

The results of Sumantadinata *et al.* (1990a) and Cherfas *et al.* (1994) relating to bimodal optima in timing of shock application are seen in other studies occasionally. Working with the African catfish *Clarias gariepinus* in Belgium, Volckaert *et al.* (1994) cited a bimodal pattern of meiotic gynogen survival in response to cold shock. Cherfas *et al.* (1993) previously described this pattern of embryo survival as reflecting alternating sensitive and non-sensitive periods during the succession of events within the developing egg.

6.3.3 Evaluating success

When various physiological shocks and their timing and duration are to be evaluated, how can researchers know if they are successful in inducing gynogenesis or androgenesis? A number of sophisticated tools are available, but one of the simplest and most elegant methods in a variety of species is the use of albino individuals and their offspring. Since albinos normally carry two copies of a recessive gene preventing pigment formation, when sperm from a normal-colored male is used for egg activation, true gynogens will be albino while pigmented offspring will indicate paternal inheritance. A similar approach can be used in androgenetic studies.

Under some practical circumstances, gynogenetic trials and their evaluation may be limited to the extent that albino females are available. Chao and Liao (1990) reported on the use of a golden-orange albino phenotype in gynogenetic experiments with the common loach *Misgurnus anguillicaudatus*, in Japan. Similarly, in Poland Goryczko *et al.* (1991) utilized a recessive gene for yellow coloration to evaluate success in inducing gynogenesis in trout. The following year, John *et al.* (1992) reported on the use of a recessive golden yellow morph in gynogenesis experiments on common carp in India. Geldhauser (1995) utilized "green" males and a golden female phenotype of the tench in gynogenesis studies in Germany. In Israel, Rothbard *et al.* (1995a) used female albino grass carp, *Ctenopharyngodon idella*, in gynogenetic studies. Eggs were activated with gamma- or UV-irradiated sperm from common carp, koi carp or normal-colored grass carp and subjected to 1–2 min heat or pressure shocks at 4–6 min post-activation ($tau_0 = 0.20$).

6.3.4 Bivalves

As mentioned previously, both the first and second polar bodies are present within a bivalve egg when it is activated. A number of methods have been used to interrupt meiosis and mitosis in bivalves including physiological shocks such as heat or pressure extremes, as well as chemical treatments such as calcium, caffeine, cytochalasin B, and 6-dimethylaminopurine (6-DMAP). The fact that chemicals (rather than physiological shocks) are often used to inhibit expulsion of polar bodies or the process of first cleavage in bivalves, coupled with the typical presence of 5N of genetic material in eggs immediately upon fertilization (2N in polar body 1 and 1N each in polar body 2, the egg nucleus, and the sperm) can lead to some unique approaches and results in bivalve chromosomal manipulation.

Recent gynogenetic work with bivalves for the most part parallels approaches developed for finfish. Cytochalasin B, a chemical treatment, seems to be the method of choice for most researchers for halting the mechanisms that result in the loss of the first and second polar bodies in bivalves. Working with *Mytilus edulis*, Fairbrother (1994) was able to produce diploid gynogens by using irradiated sperm and cytochalasin B to inhibit meiosis I (loss of the first polar body) or meiosis II (loss of the second polar body). Mitotic gynogens in the study, however, failed to develop normally and did not result in viable larvae.

Although Scarpa *et al.* (1994) examined the related species *Mytilus galloprovincialis* and also successfully produced meiotic gynogens using similar methods, a number of recent studies with bivalves illustrate the problems in trying to apply gynogenesis protocols developed for one species to other, supposedly similar species. This is at least partially due to a variety of reproductive (and, accordingly sexual) strategies found among bivalves.

6.3.5 Maternal influences

Numerous factors can influence the success of gynogenesis and androgenesis. One particularly important influence, although not immediately obvious, is the maternal source of eggs. Diter *et al.* (1993) reported pronounced maternal effects on gynogenetic yield in heat-shocked rainbow trout eggs, Oncorhynchus mykiss, irrespective of application time. Crandell *et al.* (1995), working in Alaska with pink salmon, *Oncorhynchus gorbuscha*, noted that when eggs were stripped at 1, 7, 10, and 13 days post-ovulation, survival rates of gynogenetic progeny were significantly affected not only by the source female but also through a time-of-stripping by source female interaction.

In related work, Smoker *et al.* (1995) noted that female *Oncorhynchus gorbuscha* donors varied significantly in the performance of their oocytes, but did not exhibit significant interaction with various heat-shock treatments, suggesting real maternal effects. In reporting on homozygous gynogenetic studies with the zebra fish, *Brachydanio rerio*, Hoerstgen-Schwark (1993) eloquently noted "a special sensibility of suitability of eggs derived from a particular female" to late heat or pressure shock.

Similar findings have been reported for androgenesis. In a study of androgenesis in common carp (*Cyprinus carpio*) Bongers *et al.* (1995) noted that yields of androgens depended mainly on the egg donor utilized. After 10 weeks of growout, significant differences in the portion of deformities were also attributable to donor females. Myers *et al.* (1995) indicated that differences in susceptibility to UV irradiation of eggs from different females represented a significant factor in the overall yield of androgenetic Nile tilapia (*Oreochromis niloticus*) (see Table 6.1).

6.4 ILLUSTRATIVE INVESTIGATIONS AND APPLICATIONS

6.4.1 Gynogenesis

Clearly, the optimum timing, duration and types of physiological shocks to induce gynogenesis will vary from one study to another, even when working with the same species. Nonetheless, a review of previous findings can often save valuable time and resources by providing starting values which can usually be easily adapted to specific conditions and breeding stock one may be forced to work with. Once methods are determined, gynogenesis can provide interesting insights into a variety of biological and genetic aspects of aquatic organisms.

Table 6.1 Examples of tilapia gynogenesis and androgenesis protocols from the literature.

Species	Treatment	Mitosis interruption	Time post-fertilization (minutes)	Duration (minutes)	Citation
Mitotic gynogenesis	Sperm treatment				
Oreochromis niloticus	UV: 300–310 muW cm^{-2} 2 min, at 4°C	Heat shock 41°C	27.5–30 at 28°C	3.5	Hussain et al. (1993)
Oreochromis niloticus	UV: 300–310 muW cm^{-2} 2 min, at 4°C	Pressure shock 630 kg cm^{-2}	40–50 at 28°C	2.0	Hussain et al. (1993)
Oreochromis niloticus	"UV irradiation"	Heat shock 42.5°C	22.5, 25, 27.5, 30	3 to 4	Myers et al. (1995)
Androgenesis	Egg denucleation				
Oreochromis niloticus	UV 450–720 J m^{-2} over 5–8 minutes	Heat shock 42.5°C	22.5, 25, 27.5, 30	3 to 4	Myers et al. (1995)
Oreochromis niloticus and O. aureus	UV 5940–6930 erg mm^{-2} over 54–63 seconds	Heat shock 41.6°C	27.5	5	Marengoni & Onoue (1998)

Elucidating sex determination and control

Gynogenesis has proven an invaluable tool in the study of sex determination in many finfish species as well. Komen *et al.* (1992) described perplexing results of gynogenetic studies with common carp (*Cyprinus carpio*). Meiotic gynogens proved to be all-female, reinforcing the assumption of female homogameity (two sets of the same sex determination factor, like XX) in this species. Mitotic gynogens, however, turned out to be roughly half male and intersexes, apparently homozygous for a recessive sex-influencing gene functioning independently of the major sex-determining system. In 1993, Komen and Richter reported additional results from their work, providing evidence for the existence of a "minor female sex-determining gene" in carp. Similarly, in Nile tilapia Muller-Belecke and Horstgen-Schwark (1995) reported all-female meiotic gynogens, while mitotic gynogens were approximately 35% males, again suggesting homozygous "secondary" genetic influences on phenotypic sex.

Meiotic gynogens of the silver barb *Puntius gonionotus* proved to be all-female (with a few undifferentiated exceptions) in a study reported by Pongthana *et al.* (1995), again suggesting female homogameity. In contrast, Howell *et al.* (1995) demonstrated in the same year that sex in the flatfish *Solea solea* is not determined by a simple XX-XY system, at least not with female homogameity, since meiotic gynogenesis resulted in production of both male and female progeny.

Linkage studies

As has been the case for sex determination mechanisms, gynogenetic studies can also shed light on linkage between genes. Linkage occurs when two genetic traits are not inherited independently, but are "linked" to some degree from generation to generation because they occur relatively close to each other on the same chromosome. If an individual carries two distinct forms of a gene at one locus, and two distinct forms of another gene at another locus on a single pair of homologous (matching) chromosomes, the rate at which these within-chromosome combinations break down through inheritance is influenced by the distance between the two loci, affecting how often during meiosis the chromosome strands "cross over," or break and recombine simultaneously at any given point.

With a knowledge of the specific genetic composition of broodstock over two or more generations, results of gynogenetic inheritance can be used to estimate the rates of gene-centromere (the point where homologous chromosome strands are anchored together) recombination on specific chromosomes. Arai *et al.* (1991) used gynogenesis to estimate the gene-centromere recombination rate at 11 loci in brook trout (*Salvelinus fontinalis*) and brook trout × Japanese charr (*S. leucomaenis*) hybrids.

Liu *et al.* (1993) presented results of gynogenetic studies with channel catfish (*Ictalurus punctatus*) indicating that the primary sex-determining locus in that species is linked to the glucose phosphate isomerase-B (GPI-B) gene. An analysis of the classes of offspring produced through recombination (crossing over) suggested the "X" and "Y" chromosomes in this species behave like autosomes (non-sex determining chromosomes). Guo and Allen (1995) reported on studies that allowed the mapping of several linkage groups of allozyme loci in dwarf surfclams.

In an approach based on sperm typing, Lie *et al.* (1994) introduced the use of a gynogenesis-like approach to produce partially developed haploid embryos for female-based linkage analysis. The fact that these haploid embryos often develop slowly for some period of time before perishing allows for comparatively large amounts of material for linkage studies.

Genetic and phenotypic variance

The influences of environment and genotype on the performance we observe in aquacultured stocks have already been discussed in previous chapters, and these relationships can become more complicated under conditions of gynogenesis. Sumantadinata *et al.* (1990b) described a "prominent trend" of increasing phenotypic variation in common carp, *Cyprinus carpio*, from controls to meiotic gynogens to mitotic gynogens. Similar results were reported by Taniguchi *et al.* (1990) in the ayu, *Plecoglossus altivelis*. Komen *et al.* (1992), in the carp study cited above, also reported increasing variation in body weight, gonad weight and egg size with increasing levels of homozygosity from normal diploids to meiotic gynogens to mitotic gynogens. Kim *et al.* (1993) described reductions in growth of *C. carpio* meiotic gynogens at 5 months of age compared to diploid siblings, and Na-Nakorn (1995) reported decreased survival and gonadosomatic index values for *Clarias macrocephalus* meiotic gynogens in comparison to their normal diploid female counterparts.

Taniguchi *et al.* (1990) demonstrated an important insight into this pattern, often revealed through gynogenetic studies – the influence of genetic variation and its corresponding impacts on environmental sensitivity (or adaptability) and resultant phenotypic variation. They interpreted equations developed by Kimura (1965) to illustrate the differences in phenotypic variation (1) within an inbreeding strain and (2) between individuals in an overall population of the ayu *Plecoglossus altivelis*:

$$V_P = [(1 - F)(V_G)] + V_E \quad (1)$$

$$V_P = [(1 + F)(V_G)] + V_E. \quad (2)$$

Taniguchi *et al.* (1990) equated the situation portrayed in Equation 2 with first generation gynogenetic offspring. In fact, the observed phenotypic variation in the study agreed closely with that predicted by Equation 2. Since F would be expected to be roughly 0.5 in meiotic (heterozygous) gynogens and 1 in mitotic (homozygous) gynogens, the projected impacts on phenotypic variation would be substantial.

Working with Nile tilapia, Hussain et al. (1995) produced meiotic and mitotic gynogens as well as normal diploid offspring, all from the same female. An examination of weight, length, gonadosomatic index, and morphological and meristic variables indicated the oftenseen general trend of decreasing means and increasing coefficients of variance, in the order of normal diploids: meiotic gynogens: mitotic gynogens. When viewed as a function of genetic variation (or lack thereof), the explanation for this phenomenon put forth by the authors was a reduction in developmental homeostasis and a corresponding hypersensitivity to environmental variables, as a result of increased levels of homozygosity. This ties in directly with the overdominance theory of inbreeding depression outlined in Chapter 5, and should signal a cautionary note to researchers anticipating uniform performance of gynogenetic clones.

In contrast to most reports of increasing phenotypic variation with increasing homozygosity levels, Valle et al. (1994) reported that variation decreased in homozygous clones of *P. altivelis*. In keeping with the overdominance theory of developmental homeostasis, though, these researchers reported that phenotypic variation was even further reduced in heterozygous (isogenic) ayu clones produced by crossing two homozygous lines. Working with rainbow trout, *Oncorhynchus mykiss*, Young et al. (1995) also reported greater variance in homozygous lines than in isogenic lines, but both types of fish exhibited decreased developmental homeostasis when compared to diploid counterparts.

The completely homozygous genomes of androgens and mitotic gynogens provides an opportunity to evaluate additive genetic variance from a unique perspective. Bongers et al. (1997) pointed out that in homozygous populations, additive variance doubles in comparison to the base population from which homozygotes were derived. They concluded that observed variance within homozygous families is composed of additive genetic variance and environmental variance, while the between-family variance for these groups should equal additive genetic variance. Using these relationships, they analyzed data from five full-sib families of homozygous common carp (*Cyprinus carpio*) gynogens. Heritability estimates for gonadosomatic index (GSI) at 13 months of age and percent normal larvae produced upon spawning at 19 months of age were 0.71 and 0.72, respectively.

Molluskan trials

In a study of Pacific oysters, *Crassostrea gigas*, Guo et al. (1993) produced viable gynogenetic diploids by using cytochalasin B to block the elimination of the second polar body from eggs fertilized with irradiated sperm. They also used similar treatments to block the expulsion of both polar bodies, yielding gynogenetic tetraploid (4N) offspring resulting from the combination of the (2N) first polar body, the (1N) second polar body and the (1N) egg pronucleus. These tetraploid gynogens, however, did not survive beyond 2 months post-fertilization.

The following year, Guo and Allen (1994) reported on the use of gynogenesis to investigate sex determination in the dwarf surfclam

Mulinia lateralis, again utilizing irradiated sperm and cytochalasin B to produce meiotic gynogens. In spite of low survival to metamorphosis, post-larval survival was high. At 2–3 months post-fertilization, all gynogens were determined to be female, suggesting female homogameity in this species. This conclusion was strengthened by the fact that triploid dwarf surfclams produced during the study exhibited the same sex ratio as their diploid counterparts. Mechanisms relating to triploidy will be discussed in the next chapter.

Other considerations in gynogenetic studies
Gynogenesis can be adapted for use in conjunction with other types of genetic investigations. Wiegertjes *et al.* (1995) used gynogenesis to produce homozygous progeny groups from a number of female carp (*Cyprinus carpio* L.) that had been characterized as exhibiting early/high or late/low antibody responses. Based on challenges at 6 and 12 months of age, the realized heritability for antibody response was estimated to be 0.29. The authors concluded that this approach could be used to produce experimental lines of carp with high or low antibody responses for use in immunology studies.

Unexpected complications can arise in studies of gynogenesis. Cherfas *et al.* (1995a) reported on spontaneous diploidization of maternal chromosomes in *Cyprinus carpio* eggs, noting one particular female in their experiments produced eggs exhibiting an unexpectedly high occurrence of spontaneous diploidization. Haploid oocytes, activated with biologically inert sperm, spontaneously developed in low numbers as viable diploid larvae, in the absence of any physiological shock to restore a diploid status.

6.4.2 Androgenesis

The concept of applying physiological shocks based on a scale referenced by tau_0 applies equally well to androgenesis. Rothbard *et al.* (1995b) fertilized UV-treated (2000–3000 J m^{-2}) eggs from mirror-scaled common carp (*Cyprinus carpio*) with sperm from ornamental koi. Heat shocking at 40°C for 2 min produced the highest number of androgens when applied at $tau_0 = 1.5$.

Occasionally, it may be possible to utilize irradiated eggs from one species as "hosts" for the production of androgens from a distinctly different (albeit related) species. Marengoni and Onoue (1998) described the use of Nile tilapia (*Oreochromis niloticus*) eggs in combination with blue tilapia (*Oreochromis aureus*) sperm for the successful production of androgenetic tilapia. Later in the same year, Bercsenyi *et al.* (1998) irradiated common carp (*Cyprinus carpio*) eggs and subsequently fertilized them with milt collected from three varieties of goldfish (*Carassius auratus*). They used heat shock to interrupt mitosis and produce diploid offspring. Their methods proved quite successful: no markers originating from common carp were observed (visually) in over 1500 androgens. These results were confirmed by RAPD analysis of a sample of androgens.

In those species exhibiting an XX-XY system of sex determination, androgenesis can occasionally be used to produce novel YY males (see Chapter 8). Since sperm will carry either an X or Y factor, viable androgens produced from Y-fertilized eggs will carry two Y copies. Bongers *et al.* (1999) reported on the production of a viable YY androgen in the common carp, *Cyprinus carpio*. When mated with a number of females from different genetic backgrounds, this fish produced all-male offspring. Additionally, when sperm from this male were used to produce a subsequent generation of androgens, the clones were all male (and presumably YY as well).

6.5 REFERENCES

Arai, K., Fujino, K., Sei, N., Chiba, T. & Kawamura, M. (1991) Estimating the rate of gene-centromere recombination at eleven isozyme loci in the *Salvelinus* species. *Nippon Suisan Gakkaishi*, **57** (Suppl. 6), 1043–1055.

Bercsenyi, M., Magyary, I., Urbanyi, B., Orban, L. & Horvath, L. (1998) Hatching out goldfish from common carp eggs: interspecific androgenesis between two cyprinid species. *Genome*, **41** (Suppl. 4), 573–579.

Bongers, A.B.J., Abarca, J.B., Zandieh-Doulabi, B., Eding, E.H., Komen, J. & Richter, C.J.J. (1995) Maternal influence on development of androgenetic clones of common carp, *Cyprinus carpio* L. *Aquaculture*, **137** (Suppl. 1–4), 139–147.

Bongers, A.B.J., Bovenhuis, H., Van Stokkom, A.C. *et al.* (1997) Distribution of genetic variance in gynogenetic or androgenetic families. *Aquaculture*, **153** (Suppl. 3–4), 225–238.

Bongers, A.B.J., Zandieh-Doulabi, B., Richter, C.J.J. & Komen, J. (1999) Viable androgenetic YY genotypes of common carp (*Cyprinus carpio* L.). *Journal of Heredity*, **90** (Suppl. 1), 195–198.

Chao, N.-H. & Liao, I.-C. (1990) Production of golden loach, *Misgurnus anguillicaudatus* by means of gynogenesis. In: *The Second Asian Fisheries Forum* (eds R. Hirano and I. Hanyu), pp. 535–538. Asian Fisheries Forum, Tokyo.

Cherfas, N., Gomelsky, B., Ben-Dom, N. & Hulata, G. (1995a) Evidence for the heritable nature of spontaneous diploidization in common carp, *Cyprinus carpio* L. eggs. *Aquaculture Research*, **26** (Suppl. 4), 289–292.

Cherfas, N., Hulata, G., Gomelsky, B.I., Ben-Dom, N. & Peretz, Y. (1995b) Chromosome set manipulations in the common carp *Cyprinus carpio* L. *Aquaculture*, **29** (Suppl. 1–4), 217.

Cherfas, N., Hulata, G. & Kozinski, O. (1992) Induced diploid gynogenesis and polyploidy in ornamental common carp. Experiments on accurate temperature shock timing during the second meiotic and first mitotic division in embryos. *Fisheries and Fishbreeding in Israel*, **25** (Suppl. 1), 4–15.

Cherfas, N.B., Hulata, G. & Kozinski, O. (1993) Induced diploid gynogenesis and polyploidy in ornamental (koi) carp, *Cyprinus carpio* L. 2. Timing of heat shock during the first cleavage. *Aquaculture*, **111**, 281–190.

Cherfas, N.B., Peretz, Y., Ben-Dom, N., Gomelsky, B. & Hulata, G. (1994) Induced gynogenesis and polyploidy in the ornamental (koi) carp, *Cyprinus carpio* L. 4. Comparative study on the effects of high- and low-temperature shocks. *Theoretical and Applied Genetics*, **89** (Suppl. 2–3), 193–197.

Crandell, P.A., Matsuoka, M.P. & Smoker, W.W. (1995) Effects of timing of stripping and female on the viability of gynogenetic and non-gynogenetic diploid pink salmon (*Oncorhynchus gorbuscha*). *Aquaculture*, **137** (Suppl. 1–4), 109–119.

Diter, A., Quillet, E. & Chourrout, D. (1993) Suppression of first egg mitosis induced by heat shocks in the rainbow trout. *Journal of Fish Biology*, **42** (Suppl. 5), 777–786.

Fairbrother, J.E. (1994) Viable gynogenetic diploid *Mytilus edulis* (L.) larvae produced by ultraviolet light irradiation and cytochalasin B shock. *Aquaculture*, **127** (Suppl. 1–2), 25–34.

Geldhauser, F. (1995) Gynogenesis in tench (*Tinca tinca* (L.)). *Polish Archives of Hydrobiology*, **42** (Suppl. 1–2), 141–146.

Goryczko, K., Dobosz, S., Maekinen, T. & Tomasik, L. (1991) UV-irradiation of rainbow trout sperm as a practical method for induced gynogenesis. *Journal of Applied Ichthyology*, **7** (Suppl. 3), 136–146.

Guo, X. & Allen, S.K. (1994) Sex determination and polyploid gigantism in the dwarf surfclam (*Mulinia lateralis* Say). *Genetics*, **138** (Suppl. 4), 1199–1206.

Guo, X. & Allen, S.K. (1995) The use of gynogenesis in mapping of allozyme loci in the dwarf surfclam, *Mulinia lateralis* Say. *Journal of Shellfish Research*, **14** (1), 266–267.

Guo, X., Hershberger, W.K., Cooper, K. & Chew, K.K. (1993) Artificial gynogenesis with ultraviolet light-irradiated sperm in the Pacific oyster, *Crassostrea gigas*. 1. Induction and survival. *Aquaculture*, **113** (Suppl. 3), 201–214.

Hoerstgen-Schwark, G. (1993) Production of homozygous diploid zebra fish (*Brachydanio rerio*). *Aquaculture*, **112** (Suppl. 1), 25–37.

Howell B.R., Baynes S.M. & Thompson D. (1995) Progress towards the identification of the sex-determining mechanism of the sole, *Solea solea* (L.), by the induction of diploid gynogenesis. *Aquaculture Research*, **26** (2), 135–140.

Hussain, M.G., McAndrew, B.J. & Penman, D.J. (1995) Phenotypic variation in meiotic and mitotic gynogenetic diploids of the Nile tilapia, *Oreochromis niloticus* (L.). *Aquaculture Research*, **26** (Suppl. 3), 205–212.

Hussain, M.G., Penman, D.J., McAndrew, B.J. & Johnstone, R. (1993) Suppression of first cleavage in the Nile tilapia, *Oreochromis niloticus* L. – A comparison of the relative effectiveness of pressure and heat shocks. *Aquaculture*, **111**, 263–270.

John, G., Gopalakrishnan, A., Lakra, W.S. & Barat, A. (1992) Inheritance of golden yellow colour in *Cyprinus carpio*. *Journal of the Inland Fisheries Society of India*, **24** (Suppl. 1), 8–10.

Kavumpurath, S. & Pandian, T.J. (1994) Induction of heterozygous and homozygous diploid gynogenesis in *Betta splendens* (Regan) using hydrostatic pressure. *Aquaculture and Fisheries Management*, **25** (Suppl. 2), 133–142.

Kim, E.O., Jeong, W.G. & Hue, J.S. (1993) Induction of gynogenetic diploids in common carp, *Cyprinus carpio* and their growth. *Bulletin of the National Fisheries Research and Development Agency (Korea)*, **47**, 83–91.

Kimura, M. (1965) *An Introduction to Population Genetics*, 2nd Edn. Baifuukan, Tokyo. [Cited in Taniguchi, N., Hatanaka, H. & Seki, S. (1990) Genetic variation in quantitative characters of meiotic- and mitotic-gynogenetic diploid ayu, *Plecoglossus altivelis*. *Aquaculture*, **85**, 223–233.]

Komen, J. & Richter, C.J. (1993) Sex control in carp. In: *Recent Advances in Aquaculture IV* (eds J.F. Muir & R.J. Roberts), pp. 78–86. Blackwell Scientific Publications, Oxford, UK.

Komen, J., Wiegertjes, G.F., van Ginneken, V.J.T., Eding, E.H. & Richter, C.J.J. (1992) Gynogenesis in common carp (*Cyprinus carpio* L.). 3. The effects of inbreeding on gonadal development of heterozygous and homozygous gynogenetic offspring. *Aquaculture*, **104** (Suppl. 1–2), 51–66.

Lie, Oe., Slettan, A., Lingaas, F., Olsaker, I., Hordvik, I. & Refstie, T. (1994) Haploid gynogenesis: A powerful strategy for linkage analysis in fish. *Animal Biotechnology*, **5** (Suppl. 1), 33–45.

Linhart, O., Kvasnicka, P., Flajshans, M. *et al.* (1995) Genetic studies with tench, *Tinca tinca* L.: Induced meiotic gynogenesis and sex reversal. *Aquaculture*, **132** (Suppl. 3–4), 239–251.

Liu, Q., Goudie, C.A., Simco, B.A. & Davis, K.B. (1993) Gene mapping of the sex-linked enzyme glucosephosphate isomerase-B in channel catfish. In: *From Discovery to Commercialization* (eds M. Carrillo, L. Dahle, J. Morales, P. Sorgeloos, N. Svennevig, & J. Wyban), Special Publication No. 19, European Aquaculture Society, Oostende, Belgium, p. 192.

Marengoni, N.G. & Onoue, Y. (1998) Ultraviolet-induced androgenesis in Nile tilapia, *Oreochromis niloticus* (L.), and hybrid Nile × blue tilapia, *O. aureus* (Steindachner). *Aquaculture Research*, **29** (Suppl. 5), 359–366.

Masaoka, T., Arai, K. & Suzuki, R. (1995) Production of androgenetic diploid loach *Misgurnus anguillicaudatus* from UV irradiated eggs by suppression of the first cleavage. *Fisheries Science, Tokyo*, **61** (Suppl. 4), 716–717.

Muller-Belecke, A. & Horstgen-Schwark, G. (1995) Sex determination in tilapia (*Oreochromis niloticus*), sex ratios in homozygous gynogenetic progeny and their offspring. *Aquaculture*, **137** (Suppl. 1–4), 57–65.

Myers, J.M., Penman, D.J., Basavaraju, Y. *et al.* (1995) Induction of diploid androgenetic and mitotic gynogenetic Nile tilapia (*Oreochromis niloticus* L.). *Theoretical and Applied Genetics*, **90** (Suppl. 2), 205–210.

Nagoya, H., Okamoto, H., Nakayama, I., Araki, K. & Onozato, H. (1996) Production of androgenetic diploids in amago salmon *Oncorhynchus masou ishikawae*. *Fisheries Science, Tokyo*, **62** (Suppl. 3), 380–383.

Na-Nakorn, U. (1995) Comparison of cold and heat shocks to induce diploid gynogenesis in Thai walking catfish (*Clarias macrocephalus*) and performances of gynogens. *Aquatic Living Resources/Ressources Vivantes Aquatiques*, **8** (4), 333–341.

Pongthana, N., Penman, D.J., Karnasuta, J. & McAndrew, B.J. (1995) Induced gynogenesis in the silver barb (*Puntius gonionotus* Bleeker) and evidence for female homogamety. *Aquaculture*, **135** (Suppl. 4), 267–276.

Rothbard, S., Hagani, Y., Moav, B., Shelton, W.L. & Rubinshtein, I. (1995a) Gynogenesis in albino grass carp, *Ctenopharyngodon idella* (Val.). In: *Proceedings of the Fifth International Symposium on the Reproductive Physiology of Fish* (eds F.W. Goetz & P. Thomas), p. 139, University of Texas at Austin.

Rothbard, S., Rubinshtein, I., David, L. & Shelton, W.L. (1995b) Ploidy manipulations aimed to produce androgenetic Japanese ornamental (koi) carp, *Cyprinus carpio* L. *Israeli Journal of Aquaculture/Bamidgeh*, **51** (Suppl. 1), 26–39.

Scarpa, J., Komatsu, A. & Wada, K.T. (1994) Gynogenetic induction in the mussel, *Mytilus galloprovincialis*. *Bull. Natl. Res. Inst. Aquacult. (Japan)/Yoshokukenho*, **23**, 33–41.

Smoker, W.W., Crandell, P.A. & Matsuoka, M. (1995) Second polar body retention and gynogenesis induced by thermal shocks in pink salmon, *Oncorhynchus gorbuscha* (Walbaum). *Aquaculture Research*, **26** (Suppl. 3), 213–219.

Sumantadinata, K., Taniguchi, N. & Sugama, K. (1990a) The necessary conditions and the use of ultraviolet irradiated sperm from different species to induce gynogenesis of Indonesian common carp. In: *The Second Asian Fisheries Forum* (eds R. Hirano & I. Hanyu), pp. 539–542. Asian Fisheries Forum, Tokyo.

Sumantadinata, K., Taniguchi, N. & Sugiarto, N. (1990b) Increased variance of quantitative characters in the two types of gynogenetic diploids of Indonesian common carp. *Nippon Suisan Gakkaishi*, **56** (Suppl. 12), 1979–1986.

Suwa, M., Arai, K. & Suzuki, R. (1994) Suppression of the first cleavage and cytogenetic studies on the gynogenetic loach. *Fisheries Science*, **60** (6), 673–681.

Taniguchi, N., Hatanaka, H. & Seki, S. (1990) Genetic variation in quantitative characters of meiotic- and mitotic-gynogenetic diploid ayu, *Plecoglossus altivelis*. *Aquaculture*, **85**, 223–233.

Taniguchi, N., Kijima, A., Tamura, T., Takegami, K. & Yamasaki, I. (1986) Color, growth and maturation in ploidy-manipulated fancy carp. *Aquaculture*, **57**, 321–328.

Valle, G., Taniguchi, N. & Tsujimura, A. (1994) Reduced variation of physiological traits in ayu clones, *Plecoglossus altivelis*. *Fisheries Science*, **60** (Suppl. 5), 523–526.

Volckaert, F.A.M., Galbusera, P.H.A., Hellemans, B.A.S., Van den Haute, C., Vanstaen, D. & Ollevier, F. (1994) Gynogenesis in the African catfish (*Clarias gariepinus*). 1. Induction of meiogynogenesis with thermal and pressure shocks. *Aquaculture*, **128**, 221–233.

Wiegertjes, G.F., Stet, R.J.M. & van Muiswinkel, W.B. (1995) Divergent selection for antibody production to produce standard carp (*Cyprinus carpio* L.) lines for the study of disease resistance in fish. *Aquaculture*, **137** (Suppl. 1–4), 257–262.

Young, W.P., Wheeler, P.A. & Thorgaard, G.H. (1995) Asymmetry and variability of meristic characters and spotting in isogenic lines of rainbow trout. *Aquaculture*, **137** (Suppl. 1–4), 67–76.

Chapter 7
Chromosomal Genetics II: Polyploidy

7.1 INTRODUCTION

Sexual reproduction is the norm for virtually all species aquaculturists deal with. This biological approach to reproduction has a number of advantages from an evolutionary standpoint, most notably in the ease of formation of new genotypes from generation to generation, especially through the process of meiosis. As we have previously reviewed, while most organisms typically possess two sets of chromosomes in each cell, designated as the 2N or diploid state, they produce haploid gametes that contain only 1N. In this way, as gametes combine to form new individuals the resultant organisms possess the same number and types of chromosomes as their parents.

It is this requirement to evenly divide chromosome sets during the formation of eggs and sperm that leads to the designation of 2N (two sets of chromosomes) and the associated term "diploid" as a normal state for most of the species aquaculturists work with. However, this system of dividing and recombining pairs of chromosomes could potentially be disrupted or entirely disabled if an animal possessed an odd number of chromosome sets, say 3N, that could not be equally halved. In addition, if organisms could be produced with multiples of their normal 2N karyotypes, say 4N for example, their gametes might also be of particular interest in the production of novel offspring. The production of organisms with greater than the normal number of chromosomal complements is generally referred to as the induction of polyploidy.

Natural polyploidy is implicated in the evolution of many groups of aquatic species, but for the purposes of aquaculture most research has focused on the development of techniques to produce triploid (3N) production stocks. Ultimate goals of developing triploid stocks include the production of organisms that function biologically in all respects other than gamete formation and related physiological and behavioral processes. In commercially important aquatic species, such functionally sterile stocks can be of value to improve growth rates and feed conversion efficiencies or simply to minimize potential ecological risks where exotic species, hybrids or transgenic stocks are being cultured.

7.2 THEORY

One of the original bases for the pursuit of triploidy in aquaculture was the desire to produce crops that grew continually and never slowed down to mature or reproduce. Potential benefits of triploidy include better feed conversion, higher survival (due to reduced aggression) and higher turnover within production systems. While this goal is not typically attainable in higher vertebrates due to resultant developmental abnormalities, many fish and shellfish gametes can be manipulated to produce viable triploid offspring. However, procedures commonly used to produce triploids can have profound impacts on the performance of resultant progeny, often negating the potential gains associated with functional sterility. Additionally, induction procedures which must be applied one spawn at a time are impractical for mass production of most triploid stocks. Other types of polyploids, such as tetraploids, are often produced in addition to triploids, although large-scale production of 3N organisms may still be the ultimate goal.

7.2.1 Mechanics of inducing polyploidy

How are triploids usually produced? The most direct approach involves techniques quite similar to those we have reviewed in the discussion of gynogenesis. Recall that once a sperm cell joins with an oocyte, the egg is "activated" and mechanisms are initiated that result in the physical expulsion of the second polar body. The second polar body might best be described as a (1N) bundle of genetic material representing one of the four sets of chromosomes originally associated with the formation of the egg but not destined to serve as its pronucleus.

Induction of "meiotic" triploidy involves applying the same types of thermal, pressure or chemical "shocks" to newly-fertilized eggs as meiotic gynogenesis (Table 7.1), with the resultant disruption of the mechanisms that would otherwise force the second polar body out of the egg (Fig. 7.1). This is far and away the most common approach to inducing triploidy in aquatic species and most references to triploidy involve this approach. Inheritance, by necessity, is unequal in a triploid. In contrast to gynogenesis, where sperm cannot actually fertilize eggs due to irradiation or species incompatibility, a viable sperm contributes 1N to the meiotic triploid zygote, as do both the egg pronucleus and the second polar body. In this way, three sets of chromosomes (one of paternal origin, two of maternal) combine within the triploid nucleus of the fertilized egg, and all three sets replicate with each successive cell division as the zygote begins its development.

Alternately, interploid triploids can sometimes be produced by crossing tetraploid (4N) individuals with normal diploids. Tetraploids, which typically produce 2N gametes, have been produced in a variety of aquatic organisms, again following the protocols developed for production of gynogens, but in this case mitotic gynogens (Table 7.2). In the production of tetraploids, normal (1N) eggs and sperm combine

Table 7.1 Reported optimal parameters for induction of triploidy in various aquatic species.

Species	Meiosis interruption	Time post-fertilization (minutes)	Duration (minutes)	Citation
Betta splendens	Heat shock 39°C	2	3	Kavumpurath & Pandian (1992)
Clarias fuscus	Cold shock 4–5°C	3–4 at 25–30°C	20–30	Fast (1998)
Crassostrea gigas	6-DMAP 450 µmol/l	15	10	Gerard et al. (1994a)
Ctenopharyngodon idella	Pressure shock 7000–8000 psi	4–5 at 26°C	up to 1.5	Rottman et al. (1991)
Cyprinus carpio	Heat shock 40–41°C	6	1.5–2 at 20°C	Recoubratsky et al. (1992)
Dicentrarchus labrax	Cold shock 0–2°C	5	20	Colombo et al. (1995)
Hypophthalmichthys nobilis	Pressure shock 500 atm	4	1.5	Aldridge et al. (1990)
Ictalurus punctatus	Cold shock 5°C	5 at 24–28°C	60	Wolters et al. (1981)
Odontesthes bonariensis	Cold shock 0.5–0.8°C	6 at 18.5°C	6	Struessmann et al. (1993)
Oncorhynchus mykiss	Heat shock 26.5°C	15 or 25 at 6–8°C	15	Diaz et al. (1993)
	Pressure shock 7000 psi	40 at 9.4°C	4	Rottmann et al. (1991)
Oreochromis niloticus	Cold shock 14°C	5 at 27°C	60	Kim et al. (1991)
Pagrus major	Cold shock 0°C	3	12	Sugama et al. (1992)
Perca flavescens	Heat shock 28 or 29°C	5 at 11°C	25	Malison et al. (1993a)
	Heat shock 30°C	2 at 11°C	10	Malison et al. (1993a)
	Pressure shock 9000 psi	5 at 11°C	12	Malison et al. (1993a)
Pomoxis annularis	Cold shock 5°C	5 at 20–22°C	90	Parsons (1993)
Salmo trutta	Heat shock 28°C	5	10–15	Quillet et al. (1991)
Silurus glanis	Heat shock 40.5°C	9	1	Linhart & Flajshans (1995)
Tinca tinca	Cold shock 0–2°C	2–5	35	Flajshans et al. (1993)
	Pressure shock 50–52.5 MPa	2–5	1, 2 & 5	Flajshans et al. (1993)

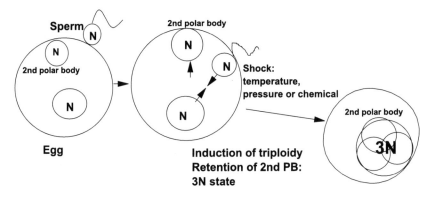

Fig. 7.1 The induction of triploidy through retention of the second polar body.

Table 7.2 Some reported optimal parameters for induction of tetraploidy.

Species	Mitosis interruption	Time post-fertilization (minutes)	Duration (minutes)	Citation
Hypophthalmichthys nobilis	Pressure shock 500 atm	36	1.5	Aldridge et al. (1990)
Perca flavescens	Pressure shock 9000 psi	192 at 11°C	24	Malison et al. (1993a)
Tinca tinca	Heat shock 40.5°C	45	1.5	Flajshans et al. (1993)
	Pressure shock 50–52.5 MPa	50	1, 2 & 5	Flajshans et al. (1993)

to form a normal, viable 2N diploid zygote. The 2N chromosomes then replicate in preparation for the first cell division, or first cleavage (Fig. 7.2), but physiological shocks are applied at the precise moment to prevent this division, leaving a 4N chromosomal complement in a single cell (Fig. 7.3). From this point on, chromosomal replication and cell division proceed normally, but each cell contains a 4N complement of chromosomes: 2N of paternal origin and 2N of maternal.

7.3 PRACTICE

In spite of frequent developmental problems during early life stages, triploidy often results in superior stocks of fish and shellfish that never seem to mature sexually and continue to grow long after their diploid counterparts have begun to divert energy to gamete formation and reproductive activities or behaviors. Heat shocks, cold shocks and pressure shocks have all been used effectively to induce meiotic triploidy in various species of fish. Chemical shocks are the typical approach with shellfish, but since the sequence of events differs slightly, we will post-

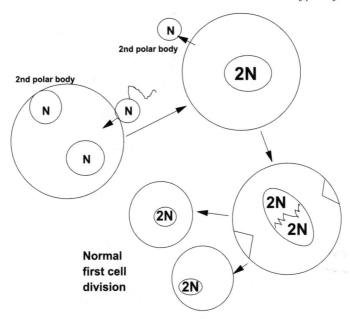

Fig. 7.2 The normal course of events during first cleavage.

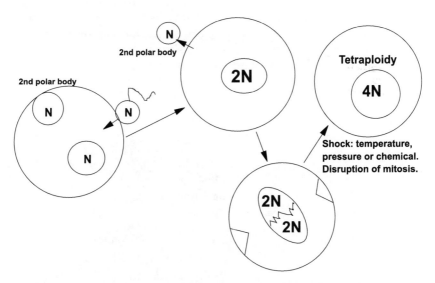

Fig. 7.3 The interruption of first cleavage, resulting in induction of tetraploidy.

pone a detailed discussion of triploidy in shellfish until after reviewing some finfish examples. Choosing which approach is most practical for ploidy manipulation in any given situation depends not only on the relative efficiency of induction methods for the species in question, but also on the numbers and general size of the eggs to be manipulated. Pressure shocks may offer a practical approach for ploidy manipulation in species with very small eggs, such as many marine fish. However,

146 *Practical Genetics for Aquaculture*

the costs associated with the large pressure cylinders and hydraulic presses that would be required to handle high numbers of relatively large eggs from species such as salmonids might become prohibitive.

7.3.1 Shock-induced versus interploid triploidy

Unfortunately, in some situations and with many species of fish, triploidy has not resulted in superior performance. Consider the thermal, pressure and chemical shocks generally utilized to force retention of the second polar body in a newly fertilized egg. Both external and internal developmental problems can result from these extreme exposures. In species that lay eggs in gelatinous masses, such as certain catfish, or in distinct strands, such as the yellow perch (*Perca flavescens*), heat shocks may degrade the associated proteins to such an extent as to cause their breakdown. The resultant colonization by opportunistic bacterial or fungal pathogens may significantly reduce survival to hatching.

Malison *et al.* (1993b) examined the internal developmental effects of heat and pressure shocks to produce meiotic triploids in yellow perch (*Perca flavescens*) and reported some very insightful observations. In one experiment, juvenile heat-shocked triploids outgrew diploids that experienced the same heat shocks but did not retain the second polar body. In another experiment using heat and pressure shocks, among shocked fish triploids again grew faster than diploids. The authors concluded that "heat and pressure shocks exert a negative influence on growth that is independent of changes in ploidy..."

Such side-effects are avoided through the production of interploid triploids (Fig. 7.4). If tetraploid (4N) animals can be produced, breed-

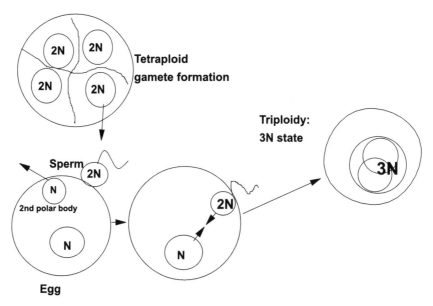

Fig. 7.4 The rationale of interploid triploidy.

ing programs can be established where either parent (male or female) contributes two sets of chromosomes in its gametes to combine with a single set from the gametes of a diploid (2N) parent, yielding mass quantities of triploid offspring without the detrimental effects of temperature or pressure shocks on subsequent development. Additionally, this approach eliminates the need to perform induction protocols on a spawn-by-spawn basis. Malison's group also conducted investigations with yellow perch in just this direction (Malison *et al.* 1993a), producing and rearing tetraploid perch to evaluate as broodstock.

The concept of producing interploid triploids in large numbers has intrigued researchers for many years. The original attraction of this approach was not necessarily to avoid the effects of post-fertilization treatments on triploid seed stock, but simply to save the effort of having to artificially produce every triploid offspring in a laboratory setting. Perhaps the only commercially viable application of meiotic triploidy to date has been the widespread use of triploid grass carp (*Ctenopharyngodon idella*) for weed control in North America and elsewhere (Fig. 7.5). Methods for inducing tetraploidy have, in theory, already been worked out during early studies of mitotic (homozygous) gynogenesis (see Chapter 6) in a number of species, the only difference being the use of normal, viable sperm to fertilize eggs destined for tetraploidy. In practice, however, the protocols developed for producing viable homozygous gynogens often require modification for successful induction of tetraploidy.

Fig. 7.5 The grass carp is widely cultured in a triploid form in North America, primarily for control of nuisance aquatic vegetation. (Photo by the author.)

In a study that tied together several relevant topics of investigation, Myers (1991) reported on meiotic- and interploid-triploid rainbow trout (*Oncorhynchus mykiss*) and their performance relative to diploid controls at the University of Washington, utilizing both inter- and intra-strain crosses. He concluded that interploid crosses appeared to be the better method of producing triploids in *O. mykiss*, and observed that interploids performed as well or better than normal diploids after ponding. While meiotic triploid groups performed poorly during the first 2 months of pond growth, some groups improved considerably during the third and fourth months of growth. Parental strains and individual broodfish within strains appeared to strongly influence the growth performance of triploid offspring, as has been the case in many other studies of triploidy. In a related study, Myers and Hershberger (1991) indicated that although tetraploid male by diploid female crosses resulted in low fertilization rates, the interploid method produced more vital fry overall than heat-shock induction.

In another example of interploid triploidy, Arai *et al.* (1993) reported on breeding trials using spontaneously occurring tetraploid loaches (*Misgurnus anguillicaudatus*). Crosses between these tetraploid males and females produced tetraploid offspring. Although concerns have occasionally been raised that production of interploid triploids should be based on using tetraploid females to avoid high nuclear to cell volume ratios, viable interploids were produced when either tetraploid females or males were crossed with diploid loaches.

7.3.2 Bivalve polyploidy

Commercially important bivalves must be included in any discussion of triploidy in aquacultured organisms. Triploid bivalves may have advantages not only in superior growth, but in the suppression of gamete formation which can negatively impact product quality from a consumer standpoint. The application of triploidy in bivalves may hold more promise than for any other group of aquacultured species, but basic biological characteristics of bivalves can create some complications in applying (and understanding) polyploidy in these mollusks.

As illustrated above, induction of triploidy in finfish generally relies on applying a temperature or pressure shock to prevent the loss of the second polar body after an egg has been fertilized and activated. To prevent confusion in previous discussions of meiosis, gynogenesis and polyploidy, the first polar body was not really discussed in detail, but as in gynogenesis it becomes an issue in the induction of triploidy in bivalves.

As mentioned previously, both the first and second polar bodies are normally present within a bivalve egg when it is fertilized, or activated. A number of methods have been used to induce polyploidy in bivalves including physiological shocks like heat or pressure extremes, as well as chemical treatments such as calcium, caffeine, cytochalasin B, and 6-dimethylaminopurine (6-DMAP). A number of polyploid states have been observed in developing bivalve embryos, but long-term viability is generally limited to triploid and tetraploid individuals.

7.4 ILLUSTRATIVE INVESTIGATIONS AND APPLICATIONS

7.4.1 Triploidy in tilapia

Tilapia may arguably be the single most appropriate candidate species for large-scale production of functionally sterile seed in the aquaculture field today. In Nile tilapia, *Oreochromis niloticus*, Kim *et al.* (1991) produced high numbers of triploids (83.3%) using a 14°C cold shock for 60 min beginning 5 min post-fertilization, to force retention of the second polar body. At 6 months of age, gonadal development was greatly reduced in triploids of both sexes. Jeong *et al.* (1992) reported optimal induction of triploidy (100%) with the use of a cold shock (10 plus or minus 0.2°C) 5 min after fertilization. Although growth of triploids and diploids was similar early in the study, by 6 months of age triploids exhibited superior growth. Gonadal examinations indicated sexual maturity in diploids, while triploids exhibited functional sterility.

Working with blue tilapia, *O. aureus*, Chang *et al.* (1993) reported distinct differences in the gonadal development of heat-shock induced triploids compared to diploids at 24 weeks of age, although growth differences were not yet apparent. Martinez-Dias and Celis-Maldonado (1994) reported 100% induction of triploidy in *O. niloticus* using both heat and cold shocks. No growth rate differences were detected up to 3 months of age, and triploid tilapia exhibited variable survival and substantial numbers of deformities (25% of cold-shocked offspring and 32% of heat-shocked offspring).

In a further examination of the effects of pressure-, heat- and cold-induced triploidy on *O. niloticus*, Hussain *et al.* (1995) reported that although triploid females were both functionally and endocrinologically sterile, triploid males exhibited endocrine profiles similar to those of normal diploid males in spite of being gametically sterile. In this study, triploids of both sexes exhibited no differences in growth or proximate composition from their diploid counterparts. In investigations of heat-shock induced *O. niloticus* meiotic triploids Bramick *et al.* (1995) also reported similar growth rates between diploids and triploids, but only up to the age of maturation. By the end of their growth trials, however, triploid males exceeded the body weights of diploid males by 66% on average (plus or minus 17%). By weight, triploid females exceeded their diploid counterparts by an average of 95% (plus or minus 27%).

7.4.2 Cyprinid results

At this point one might think the rationale behind induction of triploidy had been proven and justified based on results from the tilapia

150   *Practical Genetics for Aquaculture*

Fig. 7.6 Triploidy has often been used in hybridization studies with salmonids. Often, the results are superior to those produced by either approach independently. (Photo by the author.)

studies cited above. In spite of the commercial success of triploid grass carp (*Ctenopharyngodon idella*) (Fig. 7.6), in line with the tilapia data presented by Martinez-Dias and Celis-Maldonado (1994) and Hussain *et al.* (1995), however, triploidy results in cyprinids, salmonids and a host of other finfish have not been uniformly satisfactory. Tave (1993) reported that when cultured communally, diploid bighead carp (*Hypophthalmichthys nobilis*) significantly outgrew triploid bigheads in both length and weight during the second year of life. When cultured separately, diploids were longer and heavier than triploids (by 5% and 15%, respectively) but the differences were not statistically significant.

In a study of meiotic triploidy in common carp, *Cyprinus carpio*, Cherfas *et al.* (1994) reported, similarly, that overall survival of predominantly triploid shock-treated progeny was roughly only 70% of that observed in diploid controls by 1 year of age. Additionally, although triploids appeared to be functionally sterile their mean body weight was approximately 85% of that of diploid controls. Under every set of conditions investigated, triploid carp grew slower than their diploid siblings.

7.4.3   Salmonid polyploidy

Often, the impacts of egg shocking on triploids may linger for some time during early development, and growth advantages from sterility may not become apparent until the onset of reproductive maturity, if ever. Ihssen *et al.* (1991) outlined a textbook illustration of these phenomena with rainbow trout (*Oncorhynchus mykiss*), complete with poor triploid performance early on followed by superior growth over diploids as sexual maturation began to exert an influence. In another testimonial to the potential benefits of triploidy in salmonids, Boeuf et al. (1994) documented similar survival to smolting for triploid and diploid Atlantic salmon (*Salmo salar*), with triploid fish weighing significantly more following smoltification.

Carter et al. (1994) examined the effects of triploidy on food consumption, feeding behavior and growth in Atlantic salmon parr. When diploid and triploid parr were raised separately with their own kind, significant growth differences were noted for the first 40 days of the trial, but these differences disappeared for the remaining 52 days. In mixed groups, growth rates were not significantly different, but triploid parr appeared more likely to be recipients of agonistic behaviors. Similarly, Galbreath and Thorgaard (1995) reported that survival of all-female Atlantic salmon transferred to sea water and cultured for 376 days was 65% and 40% for communally reared diploids and triploids, respectively. Average final weight of diploids (766 g) was roughly 13% greater than that of triploids, a statistically significant difference.

In a study involving coho salmon (*Oncorhynchus kisutch*), Withler et al. (1995) found that diploids survived substantially better than triploids from fertilization to hatching (94% vs. 43%), from hatching to ponding (96% vs. 76%), during freshwater rearing (92% vs. 75%) and during marine rearing (81% vs. 60%). Smolting success was substantially greater for diploids, and diploids were 65% and 70% heavier than triploids at smolting and after two summers of marine rearing, respectively. Under these conditions and with this species, it would seem triploidy would clearly not be worth pursuing.

As a result of often lackluster results from simple hybridization, triploidy and hybridization have frequently been studied simultaneously in salmonid species (Fig. 7.7) (see Chapter 5). Oshiro et al. (1991) reported on diploid and triploid hybrids of female masu salmon (*Oncorhynchus masou*) and male Japanese charr (*Salvelinus leucomaenis pluvius*), male white-spotted charr (*Salvelinus leucomaenis leucomaenis*), and male brook trout (*Salvelinus fontinalis*). Impressively, once fry reached the swim-up stage survival of every triploid group exceeded that of its diploid counterpart.

Goryczko et al. (1992) produced triploidized reciprocal crosses of brook trout (*S. fontinalis*), sea trout (*Salmo trutta* morpha *trutta*) and rainbow trout (*O. mykiss*). Although meiotic triploidy was induced through application of heat shock, triploid hybrid survival and growth rates were comparable to those of rainbow trout controls, a result which the authors attributed to heterosis. McKay et al. (1992) documented moderate growth advantages of triploid tiger trout (*S. trutta* × *S. fontenalis* hybrids) over their diploid counterparts although this advantage was not apparent in pure brook trout, again possibly reflecting some interaction between heterosis and triploidy.

Blanc et al. (1992) examined the performance of diploid and triploid rainbow trout compared to triploid hybrids between the same rainbow trout dams and male Arctic charr (*Salvelinus alpinus*). Although

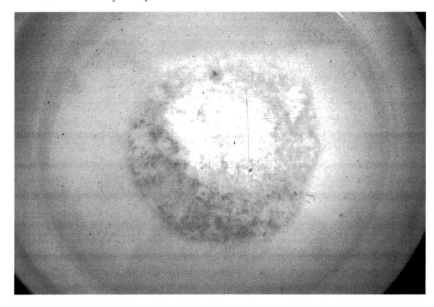

Fig. 7.7 Egg mass of channel catfish (*Ictalurus punctatus*) fertilized manually for induction of triploidy. Note how the spawn and its gelatinous matrix reflect the shape of the bowl in which fertilization and subsequent cold shock occurred. (Photo courtesy W.R. Wolters.)

survival for the triploid hybrids was comparable to that of diploid and triploid rainbow trout, growth was noticeably reduced. Even at 3 years of age and after sexual maturation had slowed the growth of the diploid rainbow trout, the triploid hybrids weighed 20% less, and were only comparable in size to rainbow trout triploids.

Galbreath and Thorgaard (1992) reported that triploid Atlantic salmon by brown trout hybrids outgrew their diploid hybrid counterparts, but did not differ significantly from diploid or triploid Atlantic salmon. In a similar salmonid study, Habicht *et al.* (1994) reported that triploid coho salmon (*Oncorhynchus kisutch*) performed comparably to diploid cohos, and better than diploid or triploid hybrids between coho females and chinook salmon (*O. tshawytscha*) males.

Diaz *et al.* (1993) provided valuable insights into subtle factors influencing the success of triploidy induction in rainbow trout. Although timing of a heat shock (26.5°C for 15 min) at 15 or 25 min post-fertilization produced similar results (73.4% and 78.6% triploidy, respectively), increased levels of induction resulted from cooler temperatures at stripping (85.9% vs. 63.0% for 6–8°C and 12.1–14°C, respectively) and from length of time that eggs remained in the body cavity (46.1%, 54.7% and 76.8% for 2, 6 and 10 days, respectively).

7.4.4 Various other finfish findings

Triploidy has been evaluated in a number of finfish species other than salmonids, cyprinids or tilapias. Wolters *et al.* (1991) presented data from growout trials of triploid channel catfish (*Ictalurus punctatus*) in earthen ponds (Fig. 7.8). When stocked at a density of 11 000 fish per hectare and fed a 32% protein floating ration, no significant differences were found between triploids and diploid controls in terms of growth or dressout percentage, but triploids exhibited lower survival and, accordingly, yield, as well as higher feed conversion ratios than diploids. An examination of gonadal development (or lack thereof) in the two groups suggested the expected functional sterility of triploids.

In the tench (*Tinca tinca*), Flajshans *et al.* (1993) used cold shocks to produce high-performing triploids, displaying much higher live weights (13.5% for females, 23.7% for males) than their diploid counterparts. Another example of positive results from the induction of triploidy in finfish, even with what would appear to be severe physiological shock to developing zygotes, is reviewed in Sugama *et al.* (1992). The authors used cold shock to induce triploidy in red sea bream (*Pagrus major*). A cold shock of 0°C applied for 12 min beginning at 3 min post-fertilization resulted in 100% triploid fish. Although early survival was lower for triploids, no differences in growth rates or overall survival between triploids and diploids was apparent after 10 months of growth. Triploids were determined to be functionally sterile, as was hoped for.

In the Sugama *et al.* (1992) study, differences in body weight detected between sea bream families through allozyme markers for diploid and triploid offspring suggested strong family influences. This serves to emphasize an important quantitative genetic aspect of triploidy – unequal inheritance. One parent (always the female in meiotic triploidy)

Fig. 7.8 Triploid grass carp (*Ctenopharyngodon idella*) await blood tests to certify ploidy level prior to sale. The most widely-produced triploid fish. (Photo courtesy T. Hymel.)

is contributing more genetic material to the offspring: roughly twice as much as the other parent. This can have implications in terms of individual broodstock or strain effects on triploid offspring. Guo et al. (1990) reported that the growth of rainbow trout (*Oncorhynchus mykiss*) triploids was significantly affected by maternal strain effects, suggesting that identification and use of specific strains and crosses within that species might lead to improved growth of triploid offspring. A similar consideration with respect to parental species contributions arises when triploid hybrids are produced.

Utilizing similar methods to investigate both meiotic gynogenesis and meiotic triploidy in the European sea bass (*Dicentrarchus labrax*), Peruzzi and Chatain (2000) indicated that induction of triploidy and gynogenesis was maximized with cold shocks of 0–1°C for a duration of 15 to 20 min, initiated at 5 min post-fertilization (or post-activation with UV-irradiated sperm, in the case of gynogenesis). Success was also attained using a 2-min pressure shock of 8500 p.s.i. beginning at 6 min post-fertilization or -activation.

The increasing interest in flatfish throughout the world has encouraged investigations in polyploidy in these species as well. Piferrer et al. (2000) reported that induction rates of roughly 90% triploidy were attained in the turbot (*Scophthalmus maximus*) using a cold shock of 0°C, applied for 20 min duration beginning at 5 min post-fertilization.

7.4.5 Bivalve studies

Perhaps one of the more unusual approaches to induction of triploidy in bivalves was reported on by Cadoret (1992). A microslide electrofusion chamber was used to test different field strengths, durations and numbers of electrical pulses on the viability of 2-cell oyster and mussel embryos. The most promising parameters were then applied to fertilized eggs in a larger chamber, producing triploids at up to 55% and 36%, and tetraploids at up to 20% and 26%, for oysters and mussels, respectively.

Scarpa et al. (1993) exposed fertilized eggs of the mussel *Mytilus galloprovincialis* to 1 mg l^{-1} of cytochalasin B from 7–35 min after fertilization. Although many treated eggs contained five pro-nuclei, presumably representing the 4N maternal contribution and the 1N paternal contribution, an examination of trocophore larvae from treated eggs suggested the presence of ploidy levels ranging from diploid to decaploid. This could potentially be explained by the period of exposure to cytochalasin B, with a decaploid larva resulting from (1) retention of the first polar body, (2) subsequent retention of the second polar body, and (3) subsequent inhibition of first cleavage in the 5N developing egg. By 1 month post-settlement, spat were 24.1%, 58.7%, and 17.2% diploid, triploid and tetraploid, respectively, suggesting that larvae with higher ploidy levels failed to survive.

Desrosiers et al. (1993) reported on the use of 6-DMAP to produce triploid Pacific oysters (*Crassostrea gigas*), giant sea scallops (*Placopecten magellanicus*) and blue mussels (*Mytilus edulis*). They produced up to

90% triploids in Pacific oysters and 95% in giant scallops, but longer exposure periods to 6-DMAP interfered with first cleavage and resulted in developmental abnormalities, especially in blue mussels. Overall, the results suggested 6-DMAP compared favorably to the traditional compound cytochalasin B, with simpler and safer protocols.

Scarpa *et al.* (1994) compared on the use of cytochalasin B, heat shocks, calcium, caffeine, and heat shocks in conjunction with calcium or caffeine to suppress the loss of the second polar body in blue mussels. Although calcium alone produced only 4.7%–7.5% triploidy, other treatments yielded fairly high rates of success: 86% for cytochalasin B, 81% for heat shock, 81% for combined heat shock and caffeine, 73% for heat shock with calcium and 71% for caffeine. Certain treatments (involving calcium and to a lesser degree caffeine) resulted in reduced developmental success of blue mussels, and at the end of 4 days development cytochalasin B was determined to be the most effective treatment in terms of production of viable triploids, with heat shock as a recommended alternative.

Gerard *et al.* (1994a) compared the use of 6-DMAP and cytochalasin B to induce triploidy in *C. gigas*. A treatment of 450 μmol l^{-1} of 6-DMAP from 15 min post-fertilization to 25 min post-fertilization resulted in 64% mean survival to D-stage larvae with 85% triploidy. In contrast, a 1 mg l^{-1} cytochalasin B treatment from 20 min post-fertilization to 35 min post-fertilization yielded a mean survival to D-stage of only 35%, with 95% triploidy. The authors cited a number of advantages of 6-DMAP over cytochalasin B, including lack of carcinogenicity, cost and water solubility. However, in their work with the Eastern oyster, *C. virginica*, Scarpa *et al.* (1995) reported disappointing results in production of viable triploids using this material.

Nell *et al.* (1994) presented results from a 30-month study evaluating the growth of triploid Sydney rock oysters (*Saccostrea commercialis*) in Australia. Sibling triploid and diploid oysters were produced for the study, to minimize genetic variation other than ploidy. Upon harvesting, triploid oysters were 41% heavier, on average, than their diploid siblings. This improvement in growth was projected to equate to an advantage of 6–18 months in time to marketable size. Triploids also exhibited higher condition factors and dry meat weights, with no differences in mortality during growout, when compared to diploids.

Sex determination in bivalves varies among species: *Crassostrea* spp. are protandric hermaphrodites, while many species are clearly dioecious (exhibiting two distinct, life-long genders). Triploidy investigations have been utilized to shed light on mechanisms of sex determination in a number of bivalve species. Guo and Allen (1994) reported on their studies of gynogenesis and triploidy in the dwarf surfclam *Mulinia lateralis*. They chose this species as a model for bivalve genetic studies due to its hardiness and early maturation. Cytochalasin B was used to block the release of the second polar body in eggs fertilized with UV-irradiated sperm (for gynogenesis) or normal sperm (for triploidy). While surviving gynogens proved to be all-female, triploids exhibited a sex ratio similar to that of diploid controls, suggesting an

XX-female XY-male mechanism of sex control. In related work published the same year (Allen & Gou 1994) the authors reported that the sex ratio of triploids and diploids was skewed in favor of males, and even more so in triploids than in diploids, suggesting possible complicating factors.

Interestingly, the triploid dwarf surfclams were significantly larger than diploid controls, but triploids had a relative fecundity of 59% for females and 80% for males when compared to their diploid counterparts. The authors hypothesized the increased body size was a result of increased triploid cell volume without a corresponding reduction in cell numbers.

The promise of interploid triploids is also being actively pursued in bivalves by several researchers. In an unusual approach to ploidy manipulation, Guo et al. (1994) reported on induction of tetraploidy in C. gigas through the use of polyethylene glycol to fuse blastomeres, with success rates of up to 30%. None of these tetraploids survived beyond the D-stage of development, probably as a by-product of treatment effects on developmental processes. In the same study, however, heat shocks were used to successfully block mitosis I (first cleavage), resulting in up to 45% tetraploid embryos.

Komaru et al. (1995a) reported on the characteristics of sperm from the tetraploid mussels (*M. galloprovincialis*) produced in the Scarpa et al. (1993) study described above. Of 16 tetraploids examined, four were found to be mature males, 11 were determined to be immature males, and one individual was classified as an hermaphrodite. Tetraploid sperm contained approximately twice the amount of DNA as diploid sperm, and were substantially larger. They also contained from five to seven mitochondria, compared to the five typically found in diploid *M. galloprovincialis* sperm (Komaru et al. 1995b). In the same year, Guo and Allen (1995) reported on the fecundity of tetraploid Pacific oysters. Like their diploid counterparts, tetraploid *C. gigas* matured at 1 year of age, with well-developed gonads and an approximately normal sex ratio. Sperm from tetraploid males had a 2N content, and eggs from tetraploid females were 70–80% larger than those from normal diploids.

A follow-up study (Guo et al. 1996) reported on the results of factorial crosses between diploid and tetraploid Pacific oysters. All crosses between diploids and tetraploids produced triploid offspring, with survival comparable to levels for diploids. A comparison treatment using cytochalasin B yielded only 46% triploids. By 8–10 months of age, interploid oysters were 13–51% larger than their diploid counterparts. And, looking to the long-term application of this approach, the authors also confirmed that tetraploid by tetraploid matings produced primarily tetraploid offspring, albeit with low survival. Readers wishing to explore this area in greater depth should refer to Guo and Allen (1997).

In follow-up work, Eudeline et al. (2000) outlined methods to increase precision in timing exposure to cytochalasin B for the production of tetraploid *Crassostrea gigas*. Eggs from triploid females were fertilized and divided into two groups, one for treatment and another for observation. Ten minutes post-fertilization, treated eggs were exposed to 0.5 mg l^{-1} cytochalasin B, while untreated eggs were monitored for extrusion of the first polar body. When 50% extrusion was judged to have occurred, cytochalasin B exposure was stopped for the treated eggs. Tetraploidy in treated eggs ranged from 13% to 92%, and at settlement the portion of tetraploid oysters averaged 45%.

7.4.6 Evaluating polyploidy induction

How do researchers know if an individual is diploid or triploid? A variety of discrimination methods has been successfully employed. In fish, a small sample of blood (approximately 1 ml) can be collected, either with a micropipette (with disposable tips, to prevent cross-contamination among specimens) or with hematocrit capillary tubes. The sample is combined with electrolyte solution and a lysing agent is added to break down blood cell membranes. The red blood cell nuclei are then examined, preferably with an electronic particle counter, to determine their volume, which can be compared to known standards developed from diploid individuals. In many cases, triploid nuclei are roughly 50% larger than their diploid counterparts. Wolters *et al.* (1982) provided details on applying this technique to determine ploidy levels in channel catfish *Ictalurus punctatus*.

Alternately, direct chromosome counts can be estimated microscopically to determine ploidy levels. Methods for producing chromosome spreads are beyond the scope of this discussion, and should be reviewed from an appropriate text on the subject or from recommendations developed by Kligerman and Bloom (1977).

Gerard *et al.* (1994b) demonstrated the application of image analysis for estimation of triploidy percentage in the Pacific oyster, *C. gigas*, Manila clam, *Ruditapes philippinarum*, and the European oyster *Ostrea edulis*. Image analysis was applied based on the optical density of stained nuclei from the individual bivalves in question. The authors demonstrated that, when using this method, the percentage of triploids in samples of *O. edulis* was consistent from the D-larval to adult stages, although different preparation techniques were required for different life stages. When comparing their method to two other common techniques for ploidy determination, Gerard *et al.* (1994b) determined that image analysis was cheaper than flow cytometry but more expensive than microfluorometry.

Child and Watkins (1994) described an approach to distinguish triploid *R. philippinarum*, based on measurements of cell nuclei from gill tissue and haemolymph. Advantages of this approach include cost of equipment (a high-power microscope and basic supplies), ease of application and minimal training requirements. This may be the best technique available, from a practical standpoint, for researchers operating with limited resources.

7.5 REFERENCES

Aldridge, F.J., Marston, R.Q. & Shireman, J.V. (1990) Induced triploids and tetraploids in bighead carp, *Hypophthalmichthys nobilis*, verified by multi-embryo cytofluorometric analysis. *Aquaculture*, **87** (Suppl. 2), 121–131.

Allen, S.K. & Guo, X. (1994) More light on sex determination in bivalves from all-female, gynogenetic *Mulinia lateralis* Say. *Journal of Shellfisheries Research*, **13** (Suppl. 1), 277.

Arai, K., Matsubara, K. & Suzuki, R. (1993) Production of polyploids and viable gynogens using spontaneously occurring tetraploid loach, *Misgurnus anguillicaudatus*. *Aquaculture*, **117** (Suppl. 3–4), 227–235.

Blanc, J.-M., Poisson, H. & Vallee, F. (1992) Survival, growth and sexual maturation of the triploid hybrid between rainbow trout and Arctic char. *Aquatic Living Resources/Ressources Vivantes Aquatiques*, **5** (Suppl. 1), 15–21.

Boeuf, G., Seddiki, H., Le Roux, A., Severe, A. & Le Bail, P.-Y. (1994) Influence of triploid status on salmon smoltification. *Aquaculture*, **121** (Suppl. 1–3), 300.

Bramick, U., Puckhaber, B., Langholz, H.-J. & Horstgen-Schwark, G. (1995) Testing of triploid tilapia (*Oreochromis niloticus*) under tropical pond conditions. *Aquaculture*, **137** (Suppl. 1–4), 343–353.

Cadoret, J.-P. (1992) Electric field-induced polyploidy in mollusc embryos. *Aquaculture*, **106** (Suppl. 2), 127–139.

Carter, C.G., McCarthy, I.D. & Houlihan, D.F. (1994) Food consumption, feeding behavior, and growth of triploid and diploid Atlantic salmon, *Salmo salar* L., parr. *Canadian Journal of Zoology*, **72** (Suppl. 4), 609–617.

Chang, S.-L., Chang, C.-F. & Liao, I.-C. (1993) Comparative study on the growth and gonadal development of diploid and triploid tilapia, *Oreochromis aureus*. *Journal of Taiwan Fisheries Research*, **1** (Suppl. 1), 43–49.

Cherfas, N.B., Gomelsky, B., Ben-Dom, N., Peretz, Y. & Hulata, G. (1994) Assessment of triploid common carp (*Cyprinus carpio* L.) for culture. *Aquaculture*, **127** (Suppl. 1), 11–18.

Child, A.R. & Watkins, H.P. (1994) A simple method to identify triploid molluscan bivalves by the measurement of cell nucleus diameter. *Aquaculture*, **125** (Suppl. 1–2), 199–204.

Colombo, L., Barbaro, A., Libertini, A., Benedetti, P., Francescon, A. & Lombardo, I. (1995) Artificial fertilization and induction of triploidy and meiogynogenesis in the European sea bass, *Dicentrarchus labrax* L. *Journal of Applied Ichthyology*, **11**, 118–125.

Desrosiers, R.R., Gerard, A., Peignon, J.-M. *et al.* (1993) A novel method to produce triploids in bivalve molluscs by the use of 6-dimethylaminopurine. *Journal of Experimental Marine Biology and Ecology*, **170** (Suppl. 1), 29–43.

Diaz, N.F., Iturra, P., Veloso, A., Estay, F. & Colihueque, N. (1993) Physiological factors affecting triploid production in rainbow trout, *Oncorhynchus mykiss*. *Aquaculture*, **114** (Suppl. 1–2), 33–40.

Eudeline, B., Allen, S.K. & Guo, X. (2000) Optimization of tetraploid induction in Pacific oysters, *Crassostrea gigas*, using first polar body as a natural indicator. *Aquaculture*, **187** (Suppl. 1–2), 73–84.

Fast, A.W. (1998) *Triploid Chinese catfish*. CTSA Publication no. 134. Center for Tropical and Subtropical Aquaculture, Makapu'u Point, Oahu, Hawaii.

Flajshans, M., Linhart, O. & Kvasnicka, P. (1993) Genetic studies of tench (*Tinca tinca* L.): Induced triploidy and tetraploidy and first performance data. *Aquaculture*, **113** (Suppl. 4), 301–312.

Galbreath, P.F. & Thorgaard, G.H. (1992) Early survival and growth rate of Atlantic salmon × brown and brook trout triploid hybrids. In: *Aquaculture '92 Book of Abstracts*, pp.98–99. World Aquaculture Society, Baton Rouge, Louisiana.

Galbreath, P.F. & Thorgaard, G.H. (1995) Saltwater performance of all-female triploid Atlantic salmon. *Aquaculture*, **138** (Suppl. 1–4), 77–85.

Gerard, A., Naciri, Y., Peignon, J.-M., Ledu, C. & Phelipot, P. (1994a) Optimization of triploid induction by the use of 6-DMAP for the oyster *Crassostrea gigas* (Thunberg). *Aquaculture and Fisheries Management*, **25** (Suppl. 7), 709–719.

Gerard, A., Naciri, Y., Peignon, J.-M. *et al.* (1994b) Image analysis: A new method for estimating triploidy in commercial bivalves. *Aquaculture and Fisheries Management*, **25** (Suppl. 7), 697–708.

Goryczko, K., Dobosz, S., Luczynski, M. & Jankun, M. (1992) Triploidized trout hybrids in aquaculture. *Bulletin of the Sea Fisheries Institute, Gdynia*, **125**, 3–6.

Guo, X. & Allen, S.K. (1994) Sex determination and polyploid gigantism in the dwarf surfclam (*Mulinia lateralis* Say). *Genetics*, **138** (Suppl. 4), 1199–1206.

Guo, X. & Allen, S.K. (1995) Tetraploid Pacific oysters (*Crassostrea gigas* Thunberg) have high fecundity despite multivalent formation during meiosis. *Journal of Shellfish Research*, **14** (Suppl. 1), 267.

Guo, X. & Allen, S.K. (1997) Sex and meiosis in autotetraploid Pacific oyster, *Crassostrea gigas* (Thunberg). *Genome*, **40** (Suppl. 3), 397–405.

Guo, X., DeBrosse, G.A. & Allen, S.K. (1996) All-triploid Pacific oysters (*Crassostrea gigas* Thunberg) produced by mating tetraploids and diploids. *Aquaculture*, **142** (Suppl. 3–4), 149–164.

Guo, X., Hershberger, W.K., Cooper, K. & Chew, K.K. (1994) Tetraploid induction with mitosis I inhibition and cell fusion in the Pacific oyster (*Crassostrea gigas* Thunberg). *Journal of Shellfish Research*, **13** (Suppl. 1), 193–198.

Guo, X., Hershberger, W.K. & Myers, J.M. (1990) Growth and survival of intrastrain and interstrain rainbow trout (*Oncorhynchus mykiss*) triploids. *Journal of the World Aquaculture Society*, **21** (Suppl. 4), 250–256.

Habicht, C., Seeb, J.E., Gates, R.B., Brock, J.R. & Olito, C.A. (1994) Triploid coho salmon outperform diploid and triploid hybrids between coho salmon and chinook salmon during their first year. *Canadian Journal of Fisheries and Aquatic Sciences*, **51** (Suppl. 1), 31–37.

Hussain, M.G., Rao, G.P.S., Humayun, N.M. *et al.* (1995) Comparative performance of growth, biochemical composition and endocrine profiles in diploid and triploid tilapia *Oreochromis niloticus* L. *Aquaculture*, **38** (Suppl. 1–4), 87–97.

Ihssen, P.E., McKay, L.R. & McMillan, I. (1991) Growth and survival of triploid rainbow trout. *Bulletin of the Aquaculture Association of Canada*, **91** (Suppl. 3), 13–15.

Jeong, W.G., Hue, J.S. & Kim, E.O. (1992) Induction of triploidy, gonadal development and growth in the Nile tilapia, *Oreochromis niloticus*. *Bulletin of the National Fisheries Research and Development Agency (Korea)*, **46**, 161–171.

Kavumpurath, S. & Pandian, T.J. (1992) Effects of induced triploidy on aggressive display in the fighting fish, *Betta splendens* Regan. *Aquaculture and Fisheries Management*, **123** (Suppl. 3), 281–290.

Kim, D.S., Choi, G.C. & Park, I, -S. (1991) Triploidy production of Nile tilapia, *Oreochromis niloticus*. *Contributions of the Institute of Marine Science, National Fisheries University of Pusan*, **23**, 235–244.

Kligerman, A.D. & Bloom, S.E. (1977) Rapid chromosome preparation from solid tissues of fish. *Journal of the Fisheries Research Board of Canada*, **34**, 266–269.

Komaru, A., Scarpa, J. & Wada, K.T. (1995a) Ultrastructure of spermatozoa in induced tetraploid mussel *Mytilus galloprovincialis* (Lmk.). *Journal of Shellfish Research*, **14** (Suppl. 2), 405–410.

Komaru, A., Wada, K.T. & Scarpa, J. (1995b) Sperm production in tetraploid mussels, *Mytilus galloprovincialis*. *Journal of Shellfish Research*, **14** (Suppl. 1), 270.

Linhart, O. & Flajshans, M. (1995) Triploidization of European catfish, *Silurus glanis* L., by heat shock. *Aquaculture Research*, **26** (Suppl. 5), 367–370.

Malison, J.A., Kayes, T.B., Held, J.A., Barry, T.P. & Amundson, C.H. (1993a) Manipulation of ploidy in yellow perch (*Perca flavescens*) by heat shock, hydrostatic pressure shock, and spermatozoa inactivation. *Aquaculture*, **110** (Suppl. 3–4), 229–242.

Malison, J.A., Procarione, L.S., Held, J.A., Kayes, T.B. & Amundson, C.H. (1993b) The influence of triploidy and heat and hydrostatic pressure shocks on the growth and reproductive development of juvenile yellow perch (*Perca flavescens*). *Aquaculture*, **116** (Suppl. 2–3), 121–133.

Martinez-Diaz, H.A. & Celis-Maldonado, C.A. (1994) Analisis de viabilidad y crecimiento haste el levante de triploides y diploides de tilapia nilotica (*Oreochromis niloticus*, Linne). *Biologia Cientifica Inpa*, **2**, 33–45.

McKay, L.R., Ihssen, P.E. & McMillan, I. (1992) Growth and mortality of diploid and triploid tiger trout (*Salmo trutta* × *Salvelinus fontinalis*). *Aquaculture*, **106** (Suppl. 3–4), 239–251.

Myers, J.M. (1991) Triploid incubation and growth performance: A comparison of meiotic and interploid triploid rainbow trout (*Oncorhynchus mykiss*) inter- and intrastrain crosses. *Dissertation Abstracts International Part B – Science and Engineering*, **51** (Suppl. 11), **DA9108503**.

Myers, J.M. & Hershberger, W.K. (1991) Early growth and survival of heat-shocked and tetraploid-derived triploid rainbow trout (*Oncorhynchus mykiss*). *Aquaculture*, **96** (Suppl. 2), 97–107.

Nell, J.A., Cox, E., Smith, I.R. & Maguire, G.B. (1994) Studies on triploid oysters in Australia. 1. The farming potential of triploid Sydney rock oysters *Saccostrea commercialis* (Iredale and Roughley). *Aquaculture*, **126** (Suppl. 3–4), 243–255.

Oshiro, T., Deng, Y., Higaki, S. & Takashima, F. (1991) Growth and survival of diploid and triploid hybrids of masu salmon *Oncorhynchus masou*. *Nippon Suisan Gakkaishi*, **57** (Suppl. 10), 1851–1857.

Parsons, G.R. (1993) Comparisons of triploid and diploid white crappies. *Transactions of the American Fisheries Society*, **122**, 237–243.

Peruzzi, S. & Chatain, B. (2000) Pressure and cold shock induction of meiotic gynogenesis and triploidy in the European sea bass, *Dicentrarchus labrax* L: relative efficiency of methods and parental variability. *Aquaculture*, **189** (Suppl. 1–2), 23–37.

Piferrer, F., Cal, R.M., Alvarez-Blazquez, B., Sanchez, L. & Martinez, P. (2000) Induction of triploidy in the turbot (*Scophthalmus maximus*) I. Ploidy determination and the effects of cold shocks. *Aquaculture*, **188** (Suppl. 1–2), 79–90.

Quillet, E., Foisil, L., Chevassus, B., Chourrout, D. & Liu, F.G. (1991) Production of all-triploid and all-female brown trout for aquaculture. *Aquatic Living Resources/Ressources Vivantes Aquatiques*, **4** (Suppl. 1), 27–32.

Recoubratsky, A.V., Gomelsky, B.I., Emelyanova, O.V. & Pankratyeva, E.V. (1992) Triploid common carp produced by heat shock with industrial fish-farm technology. *Aquaculture*, **108** (Suppl. 1–2), 13–19.

Rottmann, R.V., Shireman, J.V. & Chapman, F.A. (1991) *Induction and verification of triploidy in fish*. SRAC Publication no. 427. Southern Regional Aquaculture Center, Stoneville, Mississippi.

Scarpa, J., Toro, J.E. & Wada, K.T. (1994) Direct comparison of six methods to induce triploidy in bivalves. *Aquaculture*, **119** (Suppl. 2–3), 119–133.

Scarpa, J., Vaughan, D.E. & Longley, R. (1995) Induction of triploidy in the eastern oyster, *Crassostrea virginica*, using 6-DMAP. *Journal of Shellfish Research*, **14** (Suppl. 1), 277.

Scarpa, J., Wada, K.T. & Komaru, A. (1993) Induction of tetraploidy in mussels by suppression of polar body formation. *Nippon Suisan Gakkaishi*, **59** (Suppl. 12), 2017–2023.

Struessmann, C.A., Choon, N.B., Takashima, F. & Oshiro, T. (1993) Triploidy induction in an atherinid fish, the pejerrey (*Odontesthes bonariensis*). *The Progressive Fish-Culturist*, **55** (Suppl. 2), 83–89.

Sugama, K., Taniguchi, N., Seki, S. & Nabeshima, H. (1992) Survival, growth and gonad development of triploid red sea bream, *Pagrus major* (Timminck et Schlegel): use of allozyme markers for ploidy and family identification. *Aquaculture and Fisheries Management*, **23** (Suppl. 2), 149–159.

Tave, D. (1993) Growth of triploid and diploid bighead carp, *Hypophthalmichthys nobilis*. *Journal of Applied Aquaculture*, **2** (Suppl. 2), 13–25.

Withler, R.E., Beacham, T.D., Solar, I.I. & Donaldson, E.M. (1995) Freshwater growth, smolting, and marine survival and growth of diploid and triploid coho salmon (*Oncorhynchus kisutch*). *Aquaculture*, **136** (1–2), 91–107.

Wolters, W.R., Chrisman, C.L. & Libey, G.S. (1982) Erythrocyte nuclear measurements of diploid and triploid channel catfish, *Ictalurus punctatus*. *Journal of Fisheries Biology*, **20**, 253–258.

Wolters, W.R., Libey, G.S. & Chrisman, C.L. (1981) Induction of triploidy in channel catfish. *Transactions of the American Fisheries Society*, **110**, 312–314.

Wolters, W.R., Lilyestrom, C.G. & Craig, R.J. (1991) Growth, yield and dress-out percentage of diploid and triploid channel catfish in earthen ponds. *The Progressive Fish-Culturist*, **53** (Suppl. 1), 33–36.

Chapter 8
Sex Determination and Control

8.1   INTRODUCTION

In many cultured aquatic species, production traits such as growth rate, time or age at maturation, dressout percentage or coloration and finnage differ significantly between sexes. As a result, it is often more profitable to culture and market only the more productive or attractive sex. Even when commercial performance is not significantly influenced by gender it may be advantageous to culture monosex stocks in order to avoid unwanted or uncontrolled reproduction. Several methods of monosex culture are widely practiced, ranging from manual sorting of fingerlings to hormonal treatment of fry (Fig. 8.1). An emerging practice in many species involves phenotypic sex reversal through administration of hormones and subsequent mating of phenotypically reversed and normal broodstock to produce either monosex populations or novel individuals capable themselves of producing monosex offspring.

Two important considerations in the sex determination and control of many aquatic species are the variety of determination systems and the presence of multiple influences within some systems. There has been a tendency among aquaculture geneticists to explain mechanisms of sex determination by formulating simplistic systems with only 1, 2 or several factors to fit the results of mating trials. Unfortunately, these simplistic portraits of Mendellian-type inheritance of sex determination are often inadequate and unrealistic where aquatic species are concerned. This is already obviously clear in those species that change gender as a normal part of their life history strategies. In fact, efforts to segregate fish by sex are futile in many of these species, since a preponderance of one gender encourages conversion, or occasionally reversion, to the other.

As our understanding of molecular genetics expands, it also becomes clear that phenotypic sex in many aquatic species should be conceptualized as the end product of a complex metabolic pathway involving numerous loci and alleles, as well as environmental influences. In this sense, phenotypic sex can be viewed as a threshold trait, often influenced by major and minor genetic and environmental factors. Lester *et al.* (1989) present a well-considered review of this approach to sex determination in Nile tilapia (*Oreochromis niloticus*). Nonetheless, conceptualized systems such as XX-XY and ZZ-WZ can often provide a

Sex Determination and Control   163

Fig. 8.1 Many aquatic species, such as the tilapia fry shown here, are phenotypically undifferentiated during their first days or weeks post-hatching, allowing for the use of hormones to override their genetic sex determination and direct their development (functionally) into one gender or another. (Photo by author.)

framework sufficient for development of enlightening discussions pertaining to practical sex determination and control.

8.2   THEORY

Loci involved in sex determination may be located on specific sex chromosomes such as the X and Y chromosomes in humans or simply on chromosomes referred to as autosomes, which contain a variety of unrelated loci coding for numerous gene products. In most common conceptual systems of sex determination, sex is primarily controlled by the homozygous or heterozygous condition at one or two key loci. In turn, one sex is referred to as homogametic (producing only one type of gametes, such as X-bearing eggs) and the other as heterogametic (producing two types of gametes, such as X- or Y-bearing sperm). In spite of the genetic basis for gender, many aquatic species can be functionally directed to develop into one phenotypic sex or another during their early life stages through the administration of hormones prior to physiological differentiation. In this way, an organism that is genetically one gender can be made to function physiologically as the other. In most cases, however, these procedures result in reversal of only some individuals of the gender being altered.

The labor requirements, expense and inefficiency of hormonal sex reversal are generally not justified for commercial production of monosex stocks, and in many countries consumers may be inclined to avoid products that have been treated with hormones. These types of manipulations, however, are often potentially of great benefit on a smaller scale. The motivation for such efforts is based in the homogametic and heterogametic nature of the gender-determining or influencing loci in many aquatic species. The ultimate goal involves mass-producing single-sex stocks from novel, artificially-produced broodstock, without further need for manipulating fry.

In several commercially important crustacean species, the testis itself is not responsible for hormonal products resulting in phenotypic masculinization. This process is generally mediated by the androgenic gland. In contrast to oral or immersion administration of hormones, attempts have been made to physiologically alter phenotypic sex in freshwater decapods by masculinizing presumably homogametic females through implantation of androgenic gland tissue, with the ultimate goal of producing all-female stocks.

Hybridization can occasionally have benefits beyond simple heterosis when specific stocks can be isolated to produce all-male or all-female offspring. This is typically the result of closely related species, subspecies, or even populations within species having evolved different "systems" of sex determination. Such findings have been reported in both finfish and crustaceans. The commercial exploitation of such crosses potentially could be quite profitable, resulting not only in monosex stocks but hybrid vigor within those stocks as well.

8.3 PRACTICE

8.3.1 Homogametic monosex stocks

In many salmonid species, females are considered superior in most production traits. Since sex in salmonids is based on female homogameity, which we can refer to here as an XX state, it is possible to produce all-female stocks by mating normal females with XX "males" produced through hormonal masculinization of XX fry. Perhaps some of the earliest and most comprehensive attempts at this approach in salmonids were outlined in Johnstone et al. (1978, 1979a,b). Malison et al. (1986) explored this approach for the production of all-female yellow perch

(*Perca flavescens*). In species where males exhibit superior performance and happen to be homogametic, as in the blue tilapia *Oreochromis aureus*, this same approach can be adopted, with initial hormonal feminization rather than masculinization (Melard 1995).

Some researchers have used gynogenesis to produce all-female salmonid broods, followed by sex-reversal of these gynogenetic fry through the administration of masculinizing hormones. The final step in all-female salmonid production is to then combine milt from the sex-reversed gynogenetic fish with eggs from normal females. While functional masculinization of gynogens eliminates the need for progeny testing to determine the genetic sex of phenotypic males, inbreeding associated with gynogenesis may be a drawback in certain situations. An alternative is to simply masculinize normal salmonid fry and use progeny testing to determine which phenotypic males are actually sex-reversed XX fish.

8.3.2 Heterogametic monosex stocks

When the more desirable gender from a production standpoint is the heterogametic sex, more complex manipulations are required to allow for mass production of monosex stocks. In species where males are heterogametic, such as Nile tilapia (*Oreochromis niloticus*) and channel catfish (*Ictalurus punctatus*), it is possible to produce novel individuals homozygous for the less frequent determinant allele(s). For example, in an XX-XY system these individuals would be designated as YY. These fish can subsequently be put into service with normal (homogametic) broodstock to produce all-heterogametic offspring. Such an approach involves the sex-reversal of heterogametic individuals and their subsequent mating with their normal heterogametic counterparts. In many species the system of determination is not simple enough for this approach to work consistently or completely, as will be discussed below.

8.3.3 Minor genetic and environmental influences

Gynogenesis has proven an invaluable tool in the study of sex determination in many species (see Chapter 6). Komen *et al.* (1992) described initially perplexing results of gynogenetic studies with common carp (*Cyprinus carpio*). Although meiotic gynogens proved to be all-female, reinforcing the assumption of female homogameity in this species, mitotic gynogens turned out to be roughly half male and intersexes, apparently homozygous for a recessive sex-influencing gene functioning independently of the major sex-determining system. Komen and Richter (1993) reported additional results from their work, providing evidence for the existence of a "minor female sex-determining gene" in carp. Similar genes of "minor" influence have been documented in tilapias (see 8.4 below).

In addition to minor gender-influencing genetic factors, environmental effects should not be discounted in discussions of sex determination and control in aquatic species. Incubation and post-hatching temperatures have been implicated as sex-influencing factors in many species. Conover and Heins (1987) outlined well-documented temperature influences on sex determination in the fish *Menidia menidia*, and discussed the possible adaptive advantages of such a phenomenon. Similarly, pH has been determined to influence sex ratios in a number of freshwater fish (Rubin 1985).

8.4 ILLUSTRATIVE INVESTIGATIONS AND APPLICATIONS

8.4.1 A case study: tilapias

Hybridization
For many years, culturists have noted that certain crosses between closely related tilapia species often result in all-male production. One of the first accounts of this phenomenon was presented by Hickling (1960). Although Hickling suggested that this type of monosex production resulted from the mating of homogametic females of one species with homogametic males from another, subsequent hybridization trials produced inconsistent or contradictory results, indicating other mechanisms were involved in sex determination, at least in hybrids.

In an effort to tie together a number of diverse results from tilapia hybridization studies, Avtalion and Hammerman (1978) proposed the additional influence of a single autosomal locus, for which some species were normally homozygous dominant (*Oreochromis mossambicus* and *O. niloticus*) and others were homozygous recessive (*O. hornorum* and *O. aureus*). Although this theory represented a vast improvement in explaining results from many trials, some data still did not fit the predicted results from this model. Increasing discrepancies in hybridization results over the following years (Hulata *et al.* 1983; Majumdar & McAndrew 1983; Shelton *et al.* 1983) led many breeders and researchers to conclude that inconsistencies were due to contamination of broodstocks (Lahav & Lahav 1990) resulting from negligence or error within hatcheries, or from translocations in natural systems.

Monosex production within species
The Nile tilapia (*Oreochromis niloticus*) is almost certainly the most widely cultured species of tilapia on the planet, with numerous strains having been translocated or developed on-site throughout the globe over the past several decades. As a result of its superior qualities under most culture conditions, it has probably received more attention from researchers than all other tilapia species combined. Mair *et al.* (1991a) utilized a variety of approaches including progeny testing, gynogenesis and triploidy to investigate sex determination and control in this species.

Progeny sex ratios from apparently normal broodstock exhibited a heterogeneous, asymmetrical distribution. When a number of broods exhibiting sex ratios of 3:1 (male to female) were excluded, the distribution of sex ratios became much more homogeneous. Subsequent progeny testing of the females that produced these 3:1 broods revealed them to be heterogametic (XY), which is the normal condition for males of this species. Clearly, some environmental or genetic influence other than the simplistic XX-XY system was implicated. However, based on results of gynogenesis and triploidy trials (yielding 0:1 and 1:1 male to female progeny ratios, respectively), the authors concluded that female homogameity is the norm in this species.

In a similar in-depth study of sex determination in the blue tilapia, *Oreochromis aureus*, the same group (Mair *et al.* 1991b) examined sex ratios of single-pair matings, hybrids with other *Oreochromis* species, and progeny of sex-reversed males and females. When evaluated in terms of male:female ratios, progeny groups from normal broodstock exhibited male:female frequency peaks at 1:1, 3:5 and 1:3. Highly variable sex ratios were produced through hybridization. Sex-reversed (genetically female) males, when mated with normal females, produced progeny with sex ratios of 1:3. Similarly, most sex-reversed (genetically male) females produced all male offspring when mated to normal males. Observed deviations in sex ratios suggested the presence of an additional sex-determining locus, which apparently involved some alleles that were epistatic to those at the major sex-determining locus.

Inasmuch as administration of hormones to produce phenotypic feminization is an important step in the development of monosex tilapia populations, Potts and Phelps (1993) evaluated the effectiveness of diethylstilbestrol and ethynylestradiol as fry feeds for the production of sex-reversed Nile tilapia (*Oreochromis niloticus*). Ethynylestradiol treatments resulted in progeny groups with 57.9–65.0% females, and an additional 3.0–9.4% of fish exhibiting ovotestes. Diethylstilbestrol treatments yielded progenies with 60.3–80.0% females and 0.7–7.3% ovotestes fish. Highest success rates were attained at 400 mg kg^{-1} of feed.

In association with efforts to produce homozygous clones, Muller-Belecke and Horstgen-Schwark (1995) provided a further explanation of aberrant sex ratios and minor factors influencing phenotypic sex in Nile tilapia (*Oreochromis niloticus*). They produced 119 verified mitotic Nile tilapia gynogens, but although females of this species are typically homogametic roughly 35% of these fish were phenotypically male. When these mitotic gynogen males were mated to normal females progenies were all-male, all-female, or predominantly female. Additionally, two females that produced some mitotic (homozygous) male gynogenetic offspring produced only female meiotic (heterozygous) gynogenetic offspring. When considered altogether, these results suggest the existence of at least two minor sex-determining loci in this species, with alleles that are able to override the major XX-XY sex determination system when they occur in homozygous states.

A similar pattern of apparent accessory loci influencing phenotypic sex occasionally occurs in blue tilapia (*Oreochromis aureus*). Melard (1995) outlined methods to produce large numbers of all-male (or at least mostly male) blue tilapia based on sex reversal of normal males to phenotypic females. Melard referred to genetically male fish that exhibited female anatomy as "pseudofemales." Since males are homogametic in this species (ZZ), crossing pseudofemales with normal males would be expected to result in all male (ZZ) offspring. When *O. aureus* fry were fed 17 alpha-ethynylestradiol at 100–200 mg kg^{-1} of food for 40 days, 93–98% females resulted, compared with 53% females in control groups. As might be expected from these numbers, a little less than half the resultant females were determined to be pseudofemales. Some psuedofemales did not, however, produce all-male progeny, but fry from those that did were fed 17 alpha-ethynylestradiol at 150 mg kg^{-1} of food to produce a second generation of pseudofemales. This round of sex reversal produced highly variable results (18–95% feminization). Progeny tests on a sample of these pseudofemales indicated 97.5–100% male offspring. One notable outcome was that some of these progeny (up to 2.5%) were phenotypically female in spite of the fact that they, their parents and their "maternal" grandparents were all apparently ZZ males.

Recall the serendipitous appearance of Nile tilapia broods from XY dams with male to female ratios of 3:1 in the Mair *et al.* (1991) study cited above. Of the three males produced for every female, two would have to be XY and the third YY. Mair *et al.* (1995) compared the growth of all- or predominantly male Nile tilapia (*Oreochromis niloticus*) produced by novel YY males with hormonally sex-reversed and mixed-sex stocks of the same species. In extensively managed earthen ponds, the YY-sired stocks were significantly larger than sex-reversed fish (by 31%) and mixed-sex fish (by 59%). In fish-cum-rice and recirculating production settings, however, differences between the YY-produced fish and mixed-sex stocks (sex-reversed fish were not included) were smaller and not significant.

Subsequently, Mair *et al.* (1997) outlined their work in developing techniques for large-scale production of all-male Nile tilapia (*Oreochromis niloticus*). The initial step of this program involved the sex reversal of normal XY males to functional females through the administration of estrogen hormones. These XY "females" were then crossed to normal XY males and some portion of the resultant offspring were determined to be YY males. Progeny of YY males crossed with normal XX females were found to be 95.6% male, on average. Ultimately, broods containing YY individuals were sex reversed after hatching to produce phenotypically female YY individuals, allowing the mass production of YY male broodstock for use with normal XX females.

The occasional occurrence of phenotypic female offspring of YY males was suggested to have a genetic basis (Mair *et al.* 1997). This probably further confirms the existence of the minor sex-determining loci in tilapia hypothesized by Muller-Belecke and Horstgen-Schwark (1995) and others. Such loci could possibly be found in a homozygous state more frequently in the Mair *et al.* (1997) stocks due to inbreeding (see Chapter 5) associated with their sex-reversal and progeny-testing procedures, conducted in limited facilities. In fact, the tendency for recessive alleles to become fixed in small populations, in conjunction with the high levels of inbreeding in many hatchery stocks of tilapia, might also explain increasing inconsistencies in tilapia hybridization trials over the years as a result of fixation of "minor" sex-influencing alleles in breeding stocks at many isolated facilities.

8.4.2 Various other finfish

Shelton (1986) outlined procedures for production of all-female grass carp (*Ctenopharyngodon idella*) based on spawning masculinized females and normal females. Using a method reported by Jensen *et al.* (1983), juvenile (85–110 mm total length) female grass carp were implanted intraperitoneally with small, sealed segments of silastic (dimethylpolysiloxane) tubing packed with methyltestosterone (5 or 12 mg). Masculinized individuals produced all-female offspring when mated to normal females, suggesting female homogameity.

Malison *et al.* (1998) partially masculinized juvenile (50 mm total length) walleye (*Stizostedion vitreum*) by feeding a diet containing 17 alpha methyltestosterone (15 mg kg^{-1} feed) for a period of 60 days. At age 2, several genetic females exhibited spermatogenesis, possessing both testicular and ovarian tissue. Spermatozoa from these fish were used to fertilize eggs from normal females, and histological samples of the resultant fingerlings indicated all to be normal females.

As in so many aspects of aquaculture genetics, things do not always turn out as planned in sex-reversal initiatives. While channel catfish (*Ictalurus punctatus*) males generally grow faster and more efficiently than females, this species' response to orally administered hormonal treatments has complicated efforts to investigate its mechanisms of sex determination. In contrast to most finfish, administration of androgens such as methyl-, ethynyl- or dihydrotestosterone to sexually undifferentiated catfish fry consistently results in all-female broods. Evidence suggests that this species is characterized by female homogameity and male heterogameity, so that feminization of XY males might still be used to produce YY individuals. Unfortunately, although this approach has been pursued by several groups of researchers, unlike tilapia growers commercial channel catfish producers have realized little benefit from efforts to develop YY broodstock.

Eventually, all-male or sexually undifferentiated stocks of *I. punctatus* may be produced through administration of trenbalone acetate to undifferentiated fry (Galvez *et al.* 1995). This synthetic anabolic androgenic steroid is approved for growth enhancement of livestock in a number of countries (including the USA), and has tentatively been shown to masculinize several finfish species, including *I. punctatus*. Recent findings, however, suggest gonads of channel catfish fry exposed to this compound for extended periods may fail to develop functionally into either testes or ovaries. Such a procedure would be easily

170 *Practical Genetics for Aquaculture*

incorporated into prevailing hatchery practices, although cost:benefit relationships would require evaluation.

Finally, hybridization has been used to produce monosex stocks in a number of teleost genera other than *Oreochromis*, including the widely-recognized *Morone* and *Lepomis* complexes. A review of several investigations in this area was presented in Chapter 5.

8.4.3 Crustacean sex control and determination

In the freshwater prawn *Macrobrachium rosenbergii*, heterogenous growth among males has long been recognized as a complicating factor in harvesting and marketing (New & Singholka 1982) (Fig. 8.2). In an effort to produce all-female stocks of this commercially valuable crustacean, Nagamine *et al*. (1980) evaluated the use of surgical implantation of various male tissues in female prawns. Although females implanted with vas deferens or testicular tissues developed normally, 81% of female prawns implanted with androgenic gland (AG) tissue exhibited signs of masculinization. Sexually immature females did not develop brood chambers or reproductive setae following AG implantation, and several AG-implanted females developed male gonopore complexes. Almost all AG-implanted females developed bilaterally formed vasa deferentia, and in those females possessing male gonopores the vasa deferentia were completely formed. Two masculinized females initiated spermatogenesis in their ovarian tissue.

Fig. 8.2 The tremendous variation in growth, especially in males, associated with social hierarchies in *Macrobrachium rosenbergii* encouraged early efforts to develop all-female stocks by functionally sex-reversing genetic females into phenotypic males, in the hopes that their offspring would be all-female. (Photo courtesy J.W. Avault, Jr.)

In contrast, efforts by Nagamine and Knight (1987) to masculinize female red swamp crawfish *Procambarus clarkii* by implantation of AG tissue were less successful. The commercial appeal of all-female *P. clarkii* stocks would be most apparent in the higher dressout percentage exhibited by females. Although masculinization of pleopods was noted, no other internal or external evidence of functional sex reversal was noted.

In culture of the Australian freshwater crayfish referred to as yabbies (*Cherax* spp.), unwanted reproduction within production ponds often results in overcrowding and stunting. Some Australian growers sort through juveniles and stock only males in order to eliminate these problems and improve profitability. Interestingly, when Lawrence and Morrissy (2000) hybridized a number of freshwater crayfish subspecies within the yabby complex (*Cherax*), five crosses resulted in all-male progeny, although survival was very low in two of these crosses. Reciprocal crosses were available for comparisons with three of the five all-male producing crosses. These reciprocal crosses resulted in sex ratios of 1:1 or only a slight preponderance of males.

8.5 REFERENCES

Avtalion, R.R. & Hammerman, I.S. (1978) Sex determination in *Sarotherodon* (Tilapia). I. Introduction to a theory of autosomal influence. *Israeli Journal of Aquaculture/Bamidgeh*, **30**, 110–115.

Conover, D.O. & Heins, S.W. (1987) Adaptive variation in environmental and genetic sex determination in a fish. *Nature*, **326**, 496–498.

Galvez, J.I., Mazik, P., Phelps, R.P. & Mulvaney, D.R. (1995) Masculinization of channel catfish *Ictalurus punctatus* by oral administration of trenbalone acetate. *Journal of the World Aquaculture Society*, **26** (Suppl. 4), 378–383.

Hickling, C.F. (1960) The Malacca tilapia hybrids. *Journal of Genetics*, **57**, 1–10.

Hulata, G., Wohlfarth, G. & Rothbard, S. (1983) Progeny-testing selection of tilapia broodstocks producing all-male hybrid progenies – preliminary results. In: *Genetics in Aquaculture* (eds N.P. Wilkins & E.M. Gosling), pp. 263–268. Elsevier, Amsterdam.

Jensen, G.L., Shelton, W.L., Yang, S.L. & Wilken, L.O. (1983) Sex reversal of gynogenetic grass carp by implantation of methyltestosterone. *Transactions of the American Fisheries Society*, **112**, 79–85.

Johnstone, R., Simpson, T.H. & Walker, A.F. (1979a) Sex reversal in salmonid culture. Part III. The production and performance of all-female populations of brook trout. *Aquaculture*, **18**, 241–252.

Johnstone, R., Simpson, T.H. & Youngston, A.F. (1978) Sex reversal in salmonid culture. *Aquaculture*, **13**, 115–134.

Johnstone, R., Simpson, T.H., Youngston, A.F. & Whitehead, C. (1979b) Sex reversal in salmonid culture. Part II. The progeny of sex-reversed rainbow trout. *Aquaculture*, **18**, 13–19.

Komen, J. & Richter, C.J. (1993) Sex control in carp. In: *Recent Advances in Aquaculture IV* (eds J.F. Muir & R.J. Roberts), pp. 78–86. Blackwell Scientific Publications, Oxford, UK.

Komen, J., Wiegertjes, G.F., van Ginneken, V.J.T., Eding, E.H. & Richter, C.J.J. (1992) Gynogenesis in common carp (*Cyprinus carpio* L.). 3. The effects of inbreeding on gonadal development of heterozygous and homozygous gynogenetic offspring. *Aquaculture*, **104** (Suppl. 1–2), 51–66.

Lahav, M. & Lahav, E. (1990) The development of all-male tilapia hybrids in Nir David. *Israeli Journal of Aquaculture/Bamidgeh*, **42** (Suppl. 2), 58–61.

Lawrence, C.S. & Morrissy, N.M. (2000). Genetic improvement of marron *Cherax tenuimanus* Smith and yabbies *Cherax* spp. in Western Australia. *Aquaculture Research*, **31** (Suppl. 1), 69–82.

Lester, L.J., Lawson, K.S., Abella, T.A. & Palada, M.S. (1989) Estimated heritability of sex ratio and sexual dimorphism in tilapia. *Aquaculture and Fisheries Management*, **20**, 369–380.

Mair, C.G., Abucay, J.S., Beardmore, J.A. & Skibinski, D.O.F. (1995) Growth performance trials of genetically male tilapia (GMT) derived from YY-males in *Oreochromis niloticus* L.: On station comparisons with mixed sex and sex reversed male populations. *Aquaculture*, **137** (Suppl. 1–4), 313–323.

Mair, G.C., Abucay, J.B., Skibinski, D.O.F., Abella, T.A. & Beardmore, J.A. (1997) Genetic manipulation of sex-ratio for the large-scale production of all-male tilapia, *Oreochromis niloticus*. *Canadian Journal of Fisheries and Aquatic Sciences*, **54** (Suppl. 2), 396–404.

Mair, G.C., Scott, A.G., Penman, D.J., Beardmore, J.A. & Skibinski, D.O.F. (1991a) Sex determination in the genus *Oreochromis*. 1. Sex reversal, gynogenesis and triploidy in *O. niloticus* (L.). *Theoretical and Applied Genetics*, **82** (Suppl. 2), 144–152.

Mair, G.C., Scott, A.G., Penman, D.J., Skibinski, D.O.F. & Beardmore, J.A. (1991b) Sex determination in the genus *Oreochromis*. 2. Sex reversal, hybridisation, gynogenesis and triploidy in *O. aureus* Steindachner. *Theoretical and Applied Genetics*, **82** (Suppl. 2), 153–160.

Majumdar, K.C. & McAndrew, B.J. (1983) Sex ratios from interspecific crosses within the tilapia. In: *Proceedings of the International Symposium on Tilapia in Aquaculture* (eds L. Fishelson & Z. Yaron), pp. 261–269. Tel Aviv University, Tel Aviv, Israel.

Malison, J.A., Held, J.A., Procarione, L.S. & Garcia-Abiado, M.A.R. (1998) Production of monosex female populations of walleye from intersex broodstock. *The Progressive Fish-Culturist*, **60**, 20–24.

Malison, J.A., Kayes, T.B., Best, C.D., Amundson, C.H. & Wentworth, B.C. (1986) Sexual differentiation and the use of hormones to control sex in yellow perch (*Perca flavescens*). *Canadian Journal of Fisheries and Aquatic Sciences*, **43**, 26–35.

Melard, C. (1995) Production of a high percentage of male offspring with 17 alpha-ethynylestradiol sex-reversed *Oreochromis aureus*. 1. Estrogen sex-reversal and production of F_2 pseudofemales. *Aquaculture*, **130** (Suppl. 1), 25–34.

Muller-Belecke, A. & Horstgen-Schwark, G. (1995) Sex determination in tilapia (*Oreochromis niloticus*), sex ratios in homozygous gynogenetic progeny and their offspring. *Aquaculture*, **137** (Suppl. 1–4), 57–65.

Nagamine, C. & Knight, A.W. (1987) Masculinization of female crayfish, *Procambarus clarkii* (Girard). *International Journal of Invertebrate Reproduction and Development*, **11** (Suppl. 1), 77–88.

Nagamine, C., Knight, A.W., Maggenti, A. & Paxman, G. (1980) Masculinization of female *Macrobrachium rosenbergii* (de Man) (Decapoda, Palaemonidae) by androgenic gland implantation. *General and Comparative Endocrinology*, **41** (Suppl. 4), 442–457.

New, M.B. & Singholka, S. (1982) *Freshwater Prawn Farming. A Manual for the Culture of* Macrobrachium rosenbergii. FAO Fisheries Technical Paper No. 225, Food and Agriculture Organization of the United Nations, Rome.

Potts, A.C. & Phelps, R.P. (1993) Effectiveness of diethylstilbestrol and ethynylestradiol in the production of female Nile tilapia (*Oreochromis niloticus*) and the effect on fish morphology. In: *From Discovery to Commercialization* (eds M. Carrillo, L. Dahle, J. Morales, P. Sorgeloos, N. Svennevig & J. Wyban), p. 255. European Aquaculture Society, Oostende (Belgium).

Rubin, D.A. (1985) Effect of pH on sex ratio in cichlids and a poeciliid (Teleostei). *Copeia*, **1985**, 233–235.

Shelton, W.L. (1986) Broodstock development for monosex production of grass carp. *Aquaculture*, **57**, 311–319.

Shelton, W.L., Meriwether, F.H., Semmens, K.J. & Calhoun, W.E. (1983) Progeny sex ratios from interspecific pair spawnings of *Oreochromis* (Tilapia) *aureus* and *O. niloticus*. In: *Proceedings of the International Symposium on Tilapia in Aquaculture* (eds L. Fishelson & Z. Yaron), pp. 270–280. Tel Aviv University, Tel Aviv, Israel.

Chapter 9
Control and Induction of Maturation and Spawning

9.1 INTRODUCTION

Throughout the previous chapters we have discussed numerous approaches to genetic improvement of aquacultured species: from directed mating designs for ascertaining components of genetic variation, to selection programs, to chromosomal manipulations such as polyploidy and gynogenesis. All these approaches, however, require some ability to control, or at the very least encourage, reproduction of the species being worked with. Many procedures related to genetic research and practical improvement of aquatic species such as polyploidy, gynogenesis and complex mating designs require precise control over the entire spawning process, from arranging specific combinations of broodstock to exact timing of fertilization.

In many aquaculture situations, including genetic improvement based on simple selection programs, propagation of production stocks may require little more than good nutrition, suitable water quality, reasonably comfortable surroundings for brood animals, and the patience to allow nature to take its course. In other instances, however, it is virtually impossible under hatchery conditions to replicate the natural stimuli and circumstances that trigger some species to spawn. Unreliable captive spawning remains a major stumbling block in the development of successful commercial production systems for a variety of aquatic species.

Frequently, artificial manipulation and control of both external stimuli (through environmental control) and internal physiological processes (through hormonal intervention) is required if captive broodstocks are to be propagated and genetic manipulations are to be successfully applied (Fig. 9.1). The degree to which these techniques can be considered practical, however, often depends on the biology and life history of the species in question and the resources and expertise available.

While researchers have been quite successful over the past several decades in inducing some aquatic species to spawn, an objective examination of the numbers often suggests little advantage over natural systems in terms of the total quantity of viable offspring produced. When an advantage in sheer numbers does exist, it is usually at the expense of genetic diversity. This should be an important consideration for any restocking program utilizing hatchery-produced fry or finger-

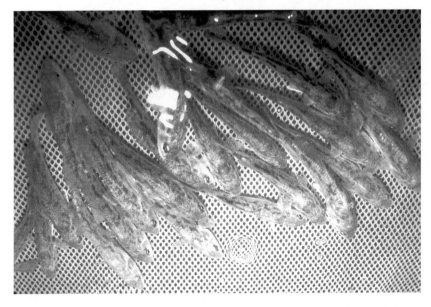

Fig. 9.1 Production of juveniles in many aquatic species (such as these hatchery-spawned red drum, *Sciaenops ocellatus*) requires the induced maturation and/or spawning of captive broodstock. (Photo by the author.)

lings to enhance natural populations. In some situations, perhaps the only apparent benefit from complicated holding and conditioning facilities is some degree of control over the timing of reproduction. In fairness, however, the only alternative to photothermal conditioning and hormonal intervention for acquiring significant numbers of eggs or fry from many aquatic species would be to procure a plankton net and head for the spawning grounds.

9.2 THEORY

9.2.1 The role of external stimuli

Most aquatic organisms have evolved to spawn under conditions that maximize the probability of survival for their offspring. Their sensory and reproductive physiology may be finely tuned to specific stimuli, or combinations of stimuli, which ultimately become essential for maturation and volitional spawning. The typical process leading to spawning involves the perception and interpretation of specific environmental conditions that, in turn, trigger a complex pathway of internal physiological developments. Unfortunately, inappropriate stimuli or physiological stressors can easily disrupt or entirely halt this chain of events.

Some species lend themselves readily to seed production in captivity, while others are virtually impossible to persuade to reproduce. Often, these characteristics reflect the range (either broad or narrow)

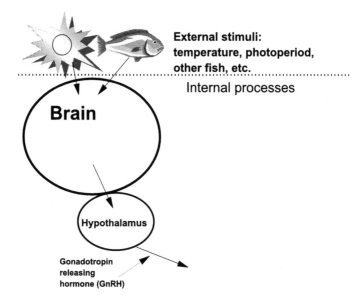

Fig. 9.2 The initial stages of the external and internal pathways to reproductive maturation.

of seasonal and environmental conditions under which the species in question would normally spawn. The relationship between external stimuli and the internal processes they trigger is not always straightforward, however. External conditions can reflect seasonal patterns, weather events, hydrologic cycles, or other environmental factors that may influence reproductive success.

Specific external stimuli that often prompt the reproductive process can include factors such as photoperiod and lunar cycles, ambient temperature or changes therein, precipitation, water flow (current), water depth, changes in barometric pressure, presence and behaviors of other fish, presence of suitable spawning substrate, and various changes in water quality, especially salinity, hardness, dissolved oxygen and pH (Fig. 9.2). Throughout nature there exists a tremendous diversity of life history strategies among aquatic species and much of this diversity reflects adaptation to take advantage of natural cycles and the opportunities for reproductive success that accompany them. If we reflect on the subtle environmental cues associated with spawning in many aquatic species, especially those with highly synchronous release of gametes, it becomes much easier to appreciate those species that spawn readily under typical culture conditions.

9.2.2 Internal processes

To reiterate, when everything works as it should, the appropriate external stimuli are usually sufficient to trigger a series of internal physiological events that result in maturation of gametes, followed by ovulation and spawning. This process can be described as a chain, the

Control and Induction of Maturation and Spawning

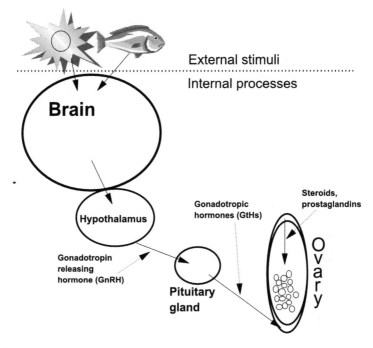

Fig. 9.3 The internal sequence of maturational stimulation and response.

links of which involve various organs and hormonal compounds. This chain, or pathway, ultimately leads from the external environment, through the brain, and all the way to the gametes themselves (Fig. 9.3). As external stimuli like temperature, photoperiod or water quality are interpreted and processed, a specific region of the brain referred to as the hypothalamus responds by producing substances such as gonadotropin releasing hormone (GnRH) and/or substances such as dopamine that inhibit gonadotropin release. The brain, the hypothalamus and the compounds it produces represent the first internal links in the reproductive pathway. The pituitary gland, typically situated directly below the brain, is stimulated by GnRH and in turn releases gonadotropic hormones (GtHs). The presence of these compounds in the bloodstream stimulates the gonads (ovaries and testes) to produce steroids that in turn trigger final egg maturation or spermiation, and prostaglandins that are involved in ovulation. When any link in the chain is broken, however, due to stress, injury, inadequate stimuli or other adverse effects, reproduction will not occur.

9.2.3 Artificial induction

Under artificial conditions, circumstances may force producers or researchers to by-pass or even override external stimuli and internal mechanisms if spawning is to successfully take place. Over the past several decades studies involving numerous species have illustrated that it is often possible to reinforce or replace specific links in the reproductive stimulus pathway through photothermal conditioning or hormonal intervention. As a general rule, the further along the pathway intervention is attempted, the greater the probability of success. A number of researchers, however, have made great advances in developing methods for induced breeding that can be applied closer and closer to the origins of the internal reproductive pathway.

9.3 PRACTICE

9.3.1 Photothermal conditioning and maturation

The environmental conditions that lead to spawning need not always be complex or subtle. Techniques for photothermal manipulation to simulate annual seasonal cycles have been well established for a number of species (Fig. 9.4). Although they may not always provide all the stimuli necessary to lead to full maturation and spawning, these manipulations are often sufficient to stimulate the development of gametes to at least an intermediate or advanced stage, often referred to as "maturation." Fish designated for seed production can be housed in tanks or aquaria and subjected to controlled temperature and photoperiod to mimic or even temporally compress annual cycles associated with maturation and spawning. These systems generally cannot be operated on a flow-through basis, requiring recirculating pumps and mechanical and biological filtration to allow continuous re-use of temperature-adjusted water (Figs 9.5, 9.6 and 9.7).

Depending on the temperature range encountered during a species' normal life history, equipment such as chillers, heaters or heat-pumps may be required to adequately manipulate water temperatures for maturation and spawning cycles (Fig. 9.8). Capacity and reliability

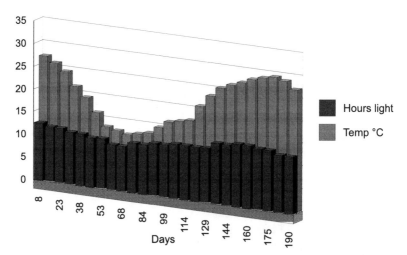

Fig. 9.4 A photoperiod-temperature regime used for captive spawning of hybrid striped bass.

Fig. 9.5 Most holding facilities for temperature and photoperiod cycling require water recirculation. Biofiltration, such as the fluidized sand bed pictured here, is essential for good water quality. Note hoses for heat transfer to or from an outside heat pump entering tank from lines along wall. Thermostat is also visible on the wall near the far corner of the room. (Photo by the author.)

are important considerations when choosing temperature manipulation equipment for outfitting photothermal conditioning facilities. Where extremely low temperatures are desired, insulation and condensation become considerations as well. Nonetheless, facilities capable of maintaining and maturing many species can be constructed with minimal investment (Figs 9.9, 9.10 and 9.11).

Photoperiod manipulation, in contrast to artificial temperature regimes, usually requires little more than conventional timers for artificial lighting and a disciplined workforce to avoid inadvertent interruptions of a photoperiod cycle once it has been established. Although temperature or photoperiod regimes alone may induce maturation or spawning in some species, in others both are essential.

9.3.2 Hormone-induced spawning

In practical terms, hormonal intervention is sometimes undertaken by choice, but more often by necessity. Holding or conditioning facilities for broodstock may be inadequate, or entirely unavailable. If wild broodstock must be collected from natural spawning grounds, hormonal intervention may be required to prevent the disruption or complete shut-down of maturation processes as a response to capture- and transport-induced stress. One key to the success of such intervention, however, is an understanding of what links of the chain are in jeopardy

Fig. 9.6 Depending on the species being spawned, an in-line chiller may be sufficient to produce cool-season temperatures. (Photo by the author.)

and what hormonal substances can be applied to reinforce or replace the natural compounds involved.

When we attempt to override or jump-start the normal physiological pathways leading to spawning, the first step in the chain, external environmental cues and conditions, may occasionally be the easiest to accommodate. Practical methods include holding mature animals in ponds under ambient conditions, or in tanks under controlled temperature and photoperiod conditions as outlined above. When this is not sufficient, however, hormones must be administered to create or reinforce internal links in the maturation chain. An appropriate example is the Australian bass *Macquaria novemaculeata*, which matures fully in fresh or brackish water ponds but does not spawn under these conditions. Battaglene and Selosse (1996) demonstrated that a single injection of hCG (100–4000 IU kg^{-1} body weight) could induce ovulation, and, frequently, natural spawning in sufficiently mature Australian bass broodstock. Hormonal intervention is usually easiest and most successful when it can be applied at later stages along the maturational pathway, and conventional wisdom suggests that hormone adminis-

Fig. 9.7 Indoor facilities for photothermal conditioning and spawning occasionally require complex mechanical infrastructure to meet the water quality and other environmental needs of the species in question. (Photo by the author.)

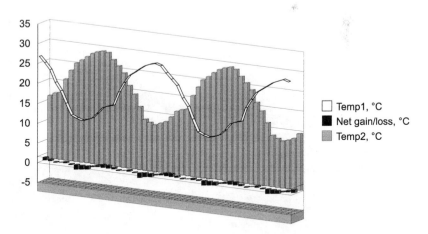

Fig. 9.8 Heat pumps can be used in conjunction with complementary thermal manipulations, resulting in minimal net temperature gain or loss.

tration should only be considered when fish are well on their way to maturation.

Gonadotropins

The most common and widely practiced intervention in the maturation pathway of fish involves administration of substitutes for the

Fig. 9.9 Induced spawning facilities can often be developed with minimal investment, provided suitable materials and equipment are used. (Photo by the author.)

Fig. 9.10 Although circular tanks are often used for photothermal conditioning, these gilthead seabream (*Sparus aurata*) spawned readily in this inexpensive rectangular tank when subjected to appropriate temperature and photoperiod regimes. (Photo by the author.)

Fig. 9.11 Another low-cost approach to induced spawning, widely used for flatfish species, involves simple outdoor pools with cloth enclosures. (Photo by the author.)

gonadotropic hormones (GtHs) normally produced by the pituitary (Fig. 9.12). These alternatives can be injected hypodermically, directly into the musculature or the peritoneal cavity to elicit similar results. One common historical approach involves utilizing actual fish pituitaries; another relies on purified gonadotropins such as human chorionic gonadotropin (hCG) to mimic the GtH that would be produced by the pituitary prior to and during the natural spawning season. In some

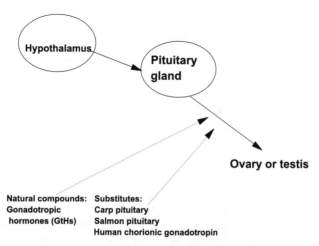

Fig. 9.12 Substitution of internally-produced gonadotropic hormones.

instances one or either of these compounds will be effective. In others, however, they must be used in combination with releasing hormones (see below) to stimulate gonads. In certain species, gonadotropins are entirely ineffective in inducing ovulation or spawning.

Pituitary extracts from species such as carp and salmon are commercially available, and can be collected from almost any variety of fish with a little effort. The technique involves sacrificing mature animals and literally removing entire pituitary glands, to be processed and injected immediately into recipient brood fish or frozen or acetone-dried for later use. To collect a fish's pituitary, the top of the skull must be removed to expose the brain cavity. The brain is then extracted from the cavity, and the pituitary gland is either separated from the base of the brain or from the bottom of the brain cavity. The pituitary tissue can be ground into a slurry and diluted to produce a supernatant for injection. The solution can be used immediately or frozen for later use.

Whole pituitaries can also be frozen for later thawing, grinding and injection. Conversely, pituitaries can be placed in acetone or absolute alcohol, which should be poured off and replaced with fresh fluid at least twice: at 1 and 10 h after collection. Some recommendations suggest a ratio, by volume, of 15 parts fluid to one part pituitaries should be maintained. After an additional 14 h in acetone or alcohol (24 h total), pituitaries can be air dried and subsequently stored in sealed jars or plastic containers for later grinding and injection. Whether fresh, frozen or dried, pituitary extracts can be highly variable in their performance depending on collection, handling and storage methods as well as maturity and health of donor fish.

Although potency can vary and effectiveness may occasionally diminish if donor and recipient species are taxonomically distant, carp and salmon pituitary injections have been used successfully on a wide variety of fish species. One occasional drawback when repeated injections are required for certain species is the occurrence of an immune or allergic response to injected pituitary extract. In spite of this, pituitary extracts from common carp often prove more suitable for spawning induction than homologous preparations produced from the species being injected. Kucharczyk *et al.* (1997), working with the bream *Abramis brama*, found that common carp pituitary produced higher levels of ovulation than bream pituitary itself when each was administered by injection in conjunction with hCG. In some instances, pituitaries from other types of animals may also produce desired results. Fagbenro *et al.* (1992) used acetone-dried pituitary extracts from the toad *Bufo regularis*, the African bull-frog *Rana adspera* and the domestic chicken *Gallus domesticus* to successfully induce ovulation and spawning in the catfish *Clarias isheriensis*.

Releasing hormones
Occasionally, events will not have proceeded to the point where GtHs can elicit a response from ovarian tissue. Conversely, certain species may simply not respond to GtHs. In these instances, hormonal intervention must target a point nearer the origin in the maturation path-

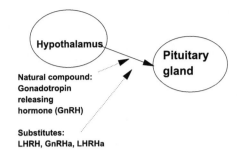

Fig. 9.13 Substitution of internally produced releasing hormone.

way. The earliest convenient point of intervention is usually the link between the hypothalamus and the pituitary (Fig. 9.13). Luteinizing-hormone-releasing hormone (LHRH) from mammals was first used to temporarily replace GnRH in fish, triggering the pituitary release of GtH and subsequent processes. In recent years, synthetic analogs of these releasing compounds have been used with much more success. These newer compounds are typically more powerful and long-lived, and easier to obtain, characterize and work with. Perhaps one of the most widespread analogs (for LHRH) is Des-GLY^{10,}[D-Ala^6]-LH-RH ethylamide, also known as LHRHa. Similarly, two of the most common analogs for GnRH are D-Lys^6-sGnRH (sGnRH-A, or salmon gonadotropin releasing hormone analog) and Des-GLY^{10,}[D-Tle^6]-GnRH (GnRHa).

Occasionally, the presence of dopamine within the bloodstream will prevent the pituitary from responding to releasing hormones or their analogs. It is often helpful, or even essential, to administer a dopamine blocker such as haloperidol, pimozide, metoclopramide or domperidone in conjunction with these releasing hormones, to prevent natural pituitary inhibition that can limit release of GtHs following stimulus from the hypothalamus. A commercial preparation combining both D-Lys^6-sGnRH and domperidone has been widely used on a variety of fish. Leu and Chou (1996) reported that administration of this commercially available solution (a single injection of 0.5 ml kg^{-1} body weight) resulted in successful spawning induction in female yellowfin porgy (*Acanthopagrus latus*), while hCG injections were entirely ineffective. Conversely, it has been demonstrated that the use of dopamine inhibitors reduces success in spawning induction in the sea bass *Dicentrarchus labrax* (Carrillo *et al.* 1995). Similar responses may be typical in a number of marine finfish (Copeland & Thomas 1989).

Steroids and prostaglandins
In some circumstances, photothermal conditioning or administration of hormones can foster the reproductive process up to an advanced stage, but the very last links in the maturation chain must be reinforced to induce ovulation and spawning. The use of serotonin to induce spawning in bivalves has been described for a number of species

186 *Practical Genetics for Aquaculture*

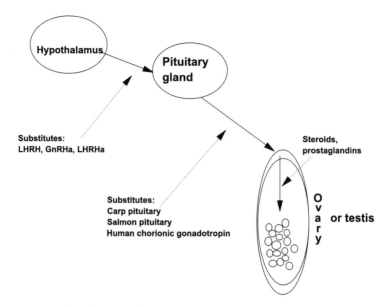

Fig. 9.14 Artificial methods for replacing an entire internal maturation pathway in order to induce maturation and spawning.

(see below). In a very few instances, purified or artificial compounds have been utilized in finfish to replicate the modes of action of steroids and prostaglandins normally produced by gonads (Fig. 9.14). One recently investigated steroidal compound is 17 alpha, 20 beta-dihydroxy-4-pregnen-3-one (known as 17,20-P), which has been implicated as the maturation-inducing steroid in walleye (*Stizostedion vitreum*) (Barry *et al*. 1995). Steroid and prostaglandin compounds have also been implicated in the expression of specific behaviors associated with spawning (Kobayashi & Stacey 1993) in finfish, which may serve as a sort of feedback to provide an external stimulus to other broodfish and reinforce the probability of successful spawning. If the previous step of the pathway involving pituitary GtHs can be sufficiently activated, however, these processes generally follow with little need for additional intervention.

9.3.3   Hormone preparation and injection practices

Perhaps one of the most important considerations in preparing, storing and administering hormones involves sanitation and sterilization procedures. Before they can be used, all materials used for mixing, storing and injection should be boiled for at least 10 min. Water or saline solutions (normally 0.7% NaCl by weight) should also be boiled briefly and cooled under sanitary conditions prior to use. Bacterial contamination of hormone solutions can result in partial or complete loss of potency over a short period of time.

If hormone solutions are going to be stored for any period of time, even under refrigeration or frozen, the addition of antibacterial compounds may be warranted. If a solution is frozen, it should be divided into appropriately sized portions for multiple administration, and clearly labeled. In this way, unneeded portions can remain frozen until they are to be used.

Dosage calculation can at times be tedious, inasmuch as the literature includes recommendations based on weight, volume and IUs (International Units). Even within a single species, recommendations may be quite variable in terms of compounds or dosage. As a result, it is often necessary to conduct range-finding trials on a small number of fish prior to deciding on a dosage appropriate for the particular species, stock and conditions encountered. As a rule, this range-finding procedure should focus on dosages slightly lower than the starting recommendation and extend to dosages roughly 1.5 times greater.

Whether based on weight, volume or IUs of hormone, almost all recommended dosages are measured and reported in terms of kg body weight of the fish to be injected. While it may be practical to obtain an actual weight for smaller fish such as tilapia, perch and certain catfish, when dealing with large broodfish it is preferable to develop a weight estimate rather than subjecting fish to the potential damage and excessive handling stress involved in direct weighing. Often, a rough length–weight relationship based on non-spawning or dead fish can be used to provide a weight estimate (Rottman *et al.* 1991a).

Once the total dosage required is determined for a specific fish, the volume of the injection becomes an important consideration. Volumes should not exceed 2 cc, even for large species, and should be divided into separate 1-cc injections in the same general region of the fish. For smaller fish, the targeted injection volume should be somewhere in the range of 0.1 to 1.0 cc. Once a recommended dose, an estimate of the fish's weight and a desired injection volume have been determined, a concentration of hormone solution can easily be calculated. The dose per kg body weight is multiplied by the body weight, and the total dose is then divide by the volume of solution desired.

Pituitary extract

To make a solution of pituitary extract, fresh, thawed or previously dried material is ground as water or saline solution is gradually added. As mentioned previously, if solution is to be stored or frozen, antibacterial compounds may be required. Bacteriostatic physiological saline, commonly used for medical or veterinary applications, is well-suited for this purpose. Once the solution has been prepared, it must be chilled to prevent breakdown of the hormonal components. While the solution is being chilled (or frozen), the pituitary tissue should settle to the bottom of the container. This material is not used in the injection, which usually is prescribed intramuscularly (see below). Only the resultant liquid solution is utilized, with the estimated potency being based on the weight of pituitary material and the volume of water or saline used. To accommodate occasional bits of tissue associated with pituitary extracts, some authors recommend using the largest gauge hypodermic needle acceptable for the species being injected. Recommended dosages normally range from 4 to 8 mg kg^{-1} of body weight. If an injection dose of 5 mg kg^{-1} body weight was to be prepared for a group of 3 kg broodfish and a desired injection volume of 1 cc had been determined, once a known weight (in mg, presumably) of pituitary material had been ground, water or saline would be added until a concentration of 15 mg ml^{-1} [(5 mg kg^{-1} × 3 kg)/1 ml] was attained.

hCG solutions
The hormone hCG is normally available through veterinary or medical suppliers in quantities of 5000 or 10 000 IUs. To prolong shelf life, this material, often in the form of a freeze-dried powder, should be stored under refrigeration at 2–7°C prior to mixing with bacteriostatic water (often supplied with the hormone in commercially available packages). Once the hormone has been mixed into solution it should be used quickly or measured into usable portions and stored in a freezer. Intramuscular administration is also the normal injection method for hCG.

Releasing hormone solutions
LHRHa generally should be handled in much the same way as hCG, but unopened material should usually be stored in a freezer. Again, bacteriostatic water should be used to dissolve and dilute the hormone, but much smaller amounts of material (measured in micrograms) of releasing hormones are usually recommended for injection. Calculations for preparing injection solutions follow the same form as with other materials: the product of dosage times fish weight, divided by the desired injection volume. Dilutions, however, may be appropriate for some dosage concentrations of releasing hormones. Care should be taken to maintain sterile materials and solutions, and unused solution should be divided into usable portions and frozen as quickly as possible. LHRHa is often administered in two or more intramuscular injections (at rates of 5–10 µg kg^{-1} body weight), so proper planning is required to prepare the required injection volumes and concentrations prior to freezing.

Other injections
Dopamine blockers and other compounds may require additional procedures prior to use. For example, haloperidol requires the addition of lactic acid to facilitate dissolution. Once this material has been added to bacteriostatic water, lactic acid should be added one drop at a time, while the solution is mixed, until the suspension becomes clear, or the pH has dropped to 3.0. Dopamine blockers are usually injected intramuscularly in conjunction with an initial injection of releasing hormone. Procedures for handling and preparation of various other hormones, dopamine blockers, steroids, prostaglandins, etc. should be determined from supplier sources.

9.3.4 Injection protocols

Although intraperitoneal injections have been recommended on occasion for certain species, when in doubt intramuscular injections are usually preferable for several reasons. An intramuscular injection into the back of the fish, either directly behind or beside the dorsal fin, generally results in less chance of injury and a more gradual and sustained uptake of the administered hormone. An intraperitoneal injection, in which the needle is inserted at the base of one of the pelvic fins, delivers compounds directly into the body cavity, but the possibility for internal injuries is greatly increased. Injections should always be administered beneath scales, when present, rather than through them.

Whether gonadotropins or releasing hormones are appropriate for the species and conditions in question, a series of two or more injections will usually produce more consistent and reliable results for females than a single injection of the total recommended dose. Males are usually injected only once, at the same time that the last, or resolving, dose is administered to the females. Recommended intervals between injections will vary by species and compound, but they usually fall somewhere in the range of 48–72 h for cold-water species, 12–24 h for cool- and warm-water species, and as little as 6–12 h for tropical species. The initial gonadotropin dose is usually 10%–33% of the total recommended dose, followed by the remaining portion of the total dose, while releasing hormones may be administered in a 20:80, 50:50 or other, 3-portion ratio. Some examples of species, dose and compound combinations reported in the literature are reviewed in Table 9.1.

9.3.5 Hormone implant technologies

Several decades ago, hormonal intervention typically relied on hypodermic injection of broodfish with whatever compound was being used, either intraperitoneally or intramuscularly. Recent research, however, has focused increasingly on the use of surgical implants that release various compounds in a more continuous and sustained pattern. Although some success has been reported using silastic or co-polymer designs, these implants are typically made of a cholesterol–cellulose matrix. Some of the earliest work in this area was described by Crim et al. (1983), Crim and Glebe (1984), Lee et al. (1986a) and Almendras et al. (1988). A review of these and other references on the topic is encouraged for those wishing to develop implant formulations and protocols. Alternately, several commercial sources of hormone implants are currently available.

The rate of hormone delivery can be adjusted by varying the ratio of cholesterol and cellulose in the implant. A 95% cholesterol content provides a relatively slow rate of release, while reducing the cholesterol content to 80% substantially increases the release rate. As is often the case with hormone injection, however, use of implants alone has not always been sufficient to produce desired results, and combination implant–injection protocols have been described for a number of species

190 *Practical Genetics for Aquaculture*

Table 9.1 Species/compound/dose combinations reported to be effective in injection-induced maturation and spawning.

Species	Compound	Dose ([kg bw]$^{-1}$ each injection)			Number injections	Injection interval (h)
		Males	Both sexes	Females		
Acanthopagrus latus	sGnRHa *and* domperidone	–		~ 50 µg ~ 5 mg	1	–
Aristichthys nobilis	hCG *and* carp pituitary	– 0.1 mg		100 & 850 IU 1.0 mg	2 1	12 after 24
Catla catla	carp pituitary *and* hCG		1.3 & 2.7 mg 167 & 333 IU		2 2	6 6
Cirrhinus mrigala	carp pituitary *and* hCG		0.7 & 1.3 mg 67 & 133 IU		2 2	6 6
Clarias fuscus	hCG	–		4000 IU	1	–
Ctenopharyngodon idella	hCG *and* carp pituitary	– 0.1 mg		100 & 850 IU 1.0 mg	2 1	12 after 24
Dicentrachus labrax	hCG LHRHa	–	800 IU	5 & 10 µg	2 2	6 4-6
Heteropneustes fossilis	sGnRHa		25 µg		1	
Hypophthalmichthys nobilis	hCG *and* carp pituitary	– 0.1 mg		100 & 850 IU 1.0 mg	2 1	12 after 24

Control and Induction of Maturation and Spawning

Labeo rohita	carp pituitary and hCG		1.3 & 2.7 mg 167 & 333 IU	2 2	6 6
Labeolabrax japonicus	hCG and chum pituitary extract		500 IU 20 mg	1	—
Lutjanus campechanus	hCG	550 IU		1	—
Lutjanus analis	hCG	500 IU	500 & 1000 IU	2 (females) 1 (males)	24 (females)
Micropterus salmoides	hCG LHRHa		4000 IU 0.5 mg	1 1	— —
Morone chrysops	hCG	145–1120 IU	200–1650 IU	1	—
Morone saxatilis	hCG	110–1100 IU	165–560 IU	1	—
Pagrus auratus	hCG		100 IU	1	—
Perca flavescens	hCG	—	150–660 IU	1	—
Siganus guttatus	hCG	—	2000 IU	2	24
Stizostedion canadense	hCG	1100 IU	1100–2200 IU	1	—
Stizostedion vitreum	hCG	165–880 IU	320–1825 IU	1–3	72

in recent years (see below). In keeping with the more practical focus of this text, however, our emphasis in this chapter will be on the use of more easily obtainable compounds that can be administered through injection.

9.3.6 Determining maturational status

Perhaps the simplest method of determining the state of maturation in many species is a visual examination of sexual characteristics such as coloration, conformation and behavior. Reproductive maturity is often easily distinguished in male fish and crustaceans with only a cursory external examination. Some examples include the formation of pearl organs on the opercula and pectoral fins of certain cyprinids or the attainment of Form I status in cambarid crayfish. In giant clams (Tridacnidae), an orange coloration of the mantle is often indicative of spawning readiness. If necessary, a sample of milt can be easily evaluated microscopically for motility (in finfish and mollusks) by combining with water under a cover slip.

When dealing with female finfish, however, and occasionally with certain mollusks, it is often advisable to sample developing oocytes to determine their progress. This can usually be done in finfish by very gently inserting an appropriately sized (in terms of both internal and external diameter) catheter tube, preferably flexible and with a smooth or rounded end, through the vent and into the ovaries. Often, a small number of eggs will be retrieved inside the tube once it is removed; if not, a syringe or suction bulb can be used to gently draw a sample into the catheter.

The eggs' average diameter and the relative positions of their nuclei can often serve as indicators of development. While egg size and appearance may be the most practical characteristic, albeit one which requires trial and error to develop a discriminating eye, some species' eggs are sufficiently clear to allow for microscopic examination of nuclear migration. As most fish eggs mature, their nuclei migrate to the end of the egg where the sperm will enter. When a sample of oocytes exhibits high numbers of off-center nuclei when back-lit, the eggs can generally be considered mature and the fish a good candidate for hormone injection.

9.3.7 Holding and handling broodstock

Perhaps the most common drawback associated with induction of spawning through photothermal conditioning and/or hormone injections is the requisite confinement, handling and associated physiological stress suffered by broodstock (see Figs 9.15 and 9.16). Successful induction of spawning requires careful handling, good nutrition and minimization of stress. Conditions of confinement, whether for conditioning or hormonal injection, deserve careful consideration. When fish are panicked respiratory rates increase dramatically, resulting in osmotic imbalance as ions are lost across gill surfaces. If fish are crowded, panicked, or held under conditions that result in increased demand for and/or reduced levels of dissolved oxygen, blood may be shunted away from internal organs to circulate immediately through more "vital" systems such as the heart, brain and musculature. This can result in the total loss of developing oocytes or spermatocytes.

Additionally, the release of stress-related compounds such as cortisol into the bloodstream during capture or handling has been implicated in the breakdown of the internal pathways that lead to final maturation, ovulation and spawning. When species that exhibit spawning-related aggression (such as many cichlids, anabantids, and certain catfish) are confined in too small an area without opportunities for females to hide or retreat, reproduction may be greatly reduced or prevented altogether.

During the period leading up to spawning, female fishes' physiological capacities are often pushed to the limit as a result of the additional metabolic demand associated with ovarian development. At this time, their ability to adjust to deficiencies in water quality variables such as ammonia, nitrites, or extremes of pH may be severely compromised as well. Handling stress under these conditions may result in mortality in some delicate species. In the sciaenid spotted seatrout, *Cynoscion nebulosus*, this problem was so severe that Thomas and Boyd (1989) developed a method for dietary administration of hormones to avoid loss of broodstock. They injected LHRHa, dissolved in acidified saline, into dead shrimp which in turn were fed to spotted seatrout broodstock to deliver 1.0–2.5 mg of hormone per kg body weight. Spawning generally followed within 30–40 h, with high rates of fertilization and hatching.

Fine-meshed, knotless netting should always be used to minimize loss of scales and protective mucus. Similarly, hands, towels and other objects coming into contact with broodfish when they are removed from the water must be damp to prevent loss of mucus. Apart from stress responses associated with holding or handling, breeding fish may be physically injured during the frequent capture and examination associated with captive spawning. Large or extremely active fish may be injured or injure themselves during collection from spawning grounds or holding ponds or tanks.

9.3.8 Obtaining gametes: finfish

The most practical approach to obtaining offspring in most aquatic species is clearly to allow volitional spawning to proceed in holding ponds or tanks. When more precise control is desired or required, however, eggs and sperm must be taken directly from broodstock. While certain circumstances may require surgical removal of eggs (as in captive paddlefish and sturgeon stocks) or milt (as in certain catfish with convoluted testes which must be removed and macerated for directed fertilization), the most common approach to collecting eggs and milt involves manual stripping of broodstock.

Although some fish such as Nile tilapia will ovulate while being held in tanks or aquaria, even in the absence of males, hand-stripping of eggs and milt is often associated with hormonal induction of spawning. Once broodfish have been collected from spawning grounds, culture facilities or conditioning ponds or tanks and injected, they should be sorted by sex and held separately in tanks for monitoring and subsequent stripping. These tanks can be partially or completely covered to minimize stress and prevent fish from jumping out. Tanks for holding and monitoring should facilitate frequent netting of individual fish for examination (Fig. 9.15). This is an important consideration, because one key to successful egg collection involves proper determination of the timing of ovulation. Once eggs have been ovulated, their quality deteriorates rapidly and fertilization must take place during a limited period of time. While this window of opportunity may be as long as several days for some salmonids, it may be as little as 15–20 min for certain warm-water species.

Female broodstock should be checked regularly for ovulation beginning several hours before the first ovulation would be expected, based on injection dosages and temperature. The appropriate interval between examinations will depend on the species, but a good rule of thumb is every 45 min for tropical species, every hour to hour-and-a-half for warm-water and temperate species, and every several hours for cool- or cold-water species. Once some experience has been gained with specific broodstock under specific conditions, efficient protocols

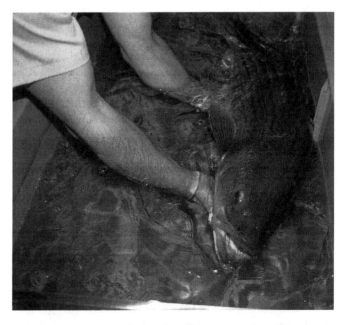

Fig. 9.15 Extra care is required when handling mature broodstock prior to strip spawning, especially for large species such as this striped bass (*Morone saxatilis*). (Photo courtesy R. Caffey.)

can be developed to maximize success while minimizing labor and handling requirements. The most common method of detecting ovulation involves physically restraining the fish in an inverted (stomach up) position and applying gentle pressure to the abdomen immediately behind the pectoral fins. As one or more fingers are stroked backward toward the vent, the expulsion of a few eggs indicates ovulation has begun. If eggs flow freely from the vent, ovulation is at or near completion, and the eggs must be taken immediately and fertilized.

Eggs must be collected without being contaminated by water or slime from the surface of the donor fish (Fig. 9.16). In eggs of many species water triggers a series of events which result in the closure of the micropyle, the portal through which the sperm enters the egg during fertilization. The fish itself, however, must be maintained under moist conditions so as to prevent the loss of the protective mucus layer from the skin if it is to be saved for recovery and future spawning. Perhaps the best approach for dealing with these simultaneous requirements is the use of a damp towel to remove excess water and slime from the female fish and subsequently cover its head and abdomen during stripping. The fish can be held for stripping, with the head slightly elevated and the vent down and slightly to one side, or placed on the edge of a counter or table on a damp towel or sheet of foam padding. In either approach, as the fish is positioned to prevent water or slime from dripping from the vent or tail the abdomen is stroked gently from front to rear to express eggs from the vent into a dry pan or bowl.

Milt can be stripped from males in much the same way as eggs are collected from females (Fig. 9.17). Male fish are blotted with a damp towel to remove any excess water, held with the head slightly elevated, and massaged to express milt onto newly stripped eggs. Enough milt should be added to cover the eggs. The contents of the bowl are then mixed thoroughly with a plastic spoon, spatula or clean feather. Once

Fig. 9.16 Taking eggs from ripe trout. Note efforts to avoid contamination with water. (Photo courtesy J.W. Avault, Jr.)

196 *Practical Genetics for Aquaculture*

Fig. 9.17 Collecting milt from from male trout. (Photo courtesy J.W. Avault, Jr.)

the eggs and milt have been thoroughly mixed, water should be added to activate the sperm. At this point, gentle stirring should continue for several minutes. Generally, only enough water to cover the eggs is required, although these proportions may vary depending on the species in question. As eggs take up water during the hardening process, however, more must be added to ensure the eggs are completely covered. In many species, after 5 to 10 min eggs should be fertilized and ready for incubation.

Storing milt
In most finfish species, milt can be collected for storage and later use. If milt is to be stored, the vent area must be well-blotted prior to milt collection. A small amount of milt should be expressed and discarded. At this point, the fish can be inverted and milt can be collected by hand stripping while drawing the fluid into a syringe. Depending on the morphology of the species being stripped, it may be convenient to insert a small plastic tube, connected to a syringe, into the urogenital pore to facilitate collection. Contamination from water, urine, blood, slime, or other sources must be avoided, and milt from individual fish should be stored separately to avoid possible cross-contamination. Milt can be stored under refrigeration in sterile plastic bags, preferably with oxygen and antibiotics. One antibiotic recommendation involves gently dissolving 50 µg of dry streptomycin sulfate per ml of milt (Rottman *et al.* 1991b).

Use of extenders for storage and application of milt is also widely described in the literature. Muiruri (1988) used an extender for tilapia milt consisting of 6.5 g l⁻¹ NaCl, 3.0 g l⁻¹ KCl, 0.2 g l⁻¹ NaHCO₃, 0.30 g l⁻¹ CaCl₂6H₂O, and 1000 ml distilled water. Harvey and Kelley (1988) reported on the use of a far more complex dilutent for storage of tilapia milt. Their mixture consisted of equal volumes of fresh hen's egg yolk and 3% sodium citrate (w/v) in distilled water. Gentamycin sulfate at a concentration of 400 µg, or 1000 IU penicillin and 800 µg streptomycin sulfate were then added per ml of solution. Finally, 5.5 mg per ml of sodium pyruvate was added to provide an energy source. The diluent was then mixed with equal volumes of milt for storage in closed vials with atmospheric oxygen. After 20 days storage under refrigeration, the suspension was allowed to stand at room temperature for 1 h, mixed with water at a 1:29 ratio, and applied to tilapia eggs. High levels of fertility were attained, with roughly 91% survival to the end of yolk sac absorption.

Adhesive eggs
In the case of species that produce adhesive eggs, special measures are taken during artificial fertilization and water hardening (Fig. 9.18). Silt-clay, bentonite and Fuller's earth have all been used to eliminate adhesiveness in fish eggs (Fig. 9.19). Saturated suspensions of these materials are used instead of water during the activation and water-hardening processes following mixing of eggs and milt. In the case of carp (*Cyprinus carpio*) eggs, the addition of water for activation often causes clumping almost immediately. An alternative solution can be made by adding 15 g of urea and 20 g of salt to 5 L of water. Once the urea and salt are dissolved, the solution can be stirred into a mixture of carp eggs and milt to trigger activation and to support water hardening. The solution should be continually added as water hardening proceeds and partially exchanged several times during the process to remove the dissolved compounds responsible for adhesion. The entire water-hardening process may take over an hour for carp eggs, at which time the fertilized eggs should be dipped very briefly into a solution of tannic acid (750 mg L⁻¹), for roughly 5 s, and immediately rinsed.

A similar protocol has been developed for eliminating the adhesiveness of white bass (*Morone chrysops*) eggs, but the urea–salt solution is applied for only the first 10 min or so of fertilization and water hardening, and a more dilute (150 mg L⁻¹) solution of tannic acid is used for a longer period (approximately 6 min). In a number of species, this more dilute tannic acid solution has been used by itself for periods of 10–12 min to eliminate adhesiveness from eggs. In catfish that produce gelatinous masses of eggs, sodium sulfite solutions (15 g L⁻¹) have been used to separate the eggs for incubation, although the pH of this solution may require adjustment to match that of the hatchery or pond water being used.

198   *Practical Genetics for Aquaculture*

Fig. 9.18 Adhesive eggs pose a number of problems during artificial fertilization and incubation. After appropriate treatment to eliminate adhesion, these white bass (*Morone chrysops*) eggs are amenable to up-flow incubation. (Photo by the author.)

9.4   ILLUSTRATIVE INVESTIGATIONS AND APPLICATIONS

9.4.1   Historical development

Although techniques are constantly being refined, the groundwork for modern approaches to photothermal conditioning and induced spawning was being laid many decades ago. Hoff *et al.* (1972) reported that although temperature and photoperiod conditioning led to maturation in the pompano (*Trachinotus carolinus*), spawning was obtained only when fish were subsequently injected with hCG and pregnant mare serum. Shortly thereafter, Lasker (1974) described successful temperature and photoperiod manipulations to induce repeated spawning in northern anchovies (*Engraulis mordax*) and croaker (*Bairdiella*

Fig. 9.19 Wild-caught red drum (*Sciaenops ocellatus*) adults readily adapt to captivity, and are easily induced to spawn in quiet surroundings with proper temperature and photoperiod cycling. (Photo courtesy W.R. Wolters.)

icistia). By this time, principles underlying the use of hormones to encourage final oocyte maturation and spawning in fish were fairly well understood. The injection of mammalian gonadotropins and carp pituitary extracts to induce spawning in channel catfish had been described by Clemens and Sneed (1957) and Sneed and Clemens (1960), respectively.

During the 1980s, photothermal manipulation was evaluated for a number of freshwater and marine species in almost every part of the globe. For example, at the Aquaculture '86 conference in Reno, Nevada, reports were presented covering photothermal maturation of smallmouth bass (*Micropterus dolomieui*) in Canada (Cantin 1986) and Chinese catfish (*Clarias fuscus*) (Young *et al.* 1986) and milkfish (*Chanos chanos*) in Hawaii (Lee *et al.* 1986b). Although Cantin (1986) reported successful reproduction utilizing only temperature and photoperiod as external stimuli, Young *et al.* (1986) and Lee *et al.* (1986b) both indicated that administration of hormones (human chorionic gonadotropin [hCG] and luteinizing hormone releasing hormone analog [LHRHa], respectively) was required to complete internal processes leading up to spawning. Many studies over the years, including Hoff *et al.* (1972), have reported similar results, where external stimuli are sufficient to promote gamete maturation, but hormonal intervention is required to complete the physiological processes leading to spawning.

200 *Practical Genetics for Aquaculture*

9.4.2 A case study: the red drum (*Sciaenops ocellatus*)

In the US, more resources have probably been devoted to induction of spawning in the red drum (*Sciaenops ocellatus*) than any other species (Fig. 9.20), although various flounders have become more prominent in recent literature in this area. Arnold *et al.* (1976) reported on some of the pioneering use of photoperiod and temperature conditioning for laboratory spawning of red drum and southern flounder (*Paralichthys lethostigma*). When held in 30 000-l tanks and subjected to photoperiod and temperature regimes that compressed an annual seasonal cycle (beginning and ending in late fall) into roughly 7 months, red drum spawned voluntarily, producing eggs with a fertilization rate exceeding 99%. The largest (>2 kg) flounder females in the study also spawned, producing eggs over a 13-day period with fertilization rates ranging from 30% to 50%.

Over the following decade, a number of North American authors described facilities and systems for photothermal conditioning of red

Fig. 9.20 An earth-coupled heat pump configured for heating and cooling broodfish holding tanks. (Photo by the author.)

Control and Induction of Maturation and Spawning   201

drum. Lawson *et al.* (1989) outlined a heat-pump driven temperature exchange system for spawning red drum that would be easily adaptable for other marine species (Fig. 9.21). McCarty (1990) subsequently described in detail the design and operation of a successful red drum photoperiod/temperature spawning facility operated by the Texas Parks and Wildlife Department for stock enhancement.

Roberts (1990) reviewed the principles of using exogenous stimuli to trigger maturation and spawning in captive fish, with an emphasis on red drum. This work also included a review of practical approaches to construction and management of photoperiod/temperature spawning tank facilities, and methods and procedures used by three pioneering hatcheries producing mass quantities of red drum at that time. More recently, Falls *et al.* (1997) and Jenkins *et al.* (1997) provided new insights into the operation of red drum spawning facilities to maximize efficiency and output.

Occasionally, in the practice of spawning induction, limited facilities and numbers of broodstock can nonetheless produce phenomenal numbers of eggs and fry. Jenkins *et al.* (1997) illustrated this point

Fig. 9.21 The buoyancy of eggs from a number of finfish facilitates their collection from spawning tanks by skimming devices. (Photo courtesy W.R. Wolters.)

elegantly in their paper on control and management of captive spawning in red drum, presented at the World Aquaculture '97 conference in Seattle. Induced spawning of the red drum (*Sciaenops ocellatus*) had been practiced more or less successfully for roughly 20 years, with most hatcheries clinging to a tried and true approach. As is often the case in fish culture, however, researchers had failed to develop methods to make the most of an investment in red drum broodstock, conditioning tank facilities, and time spent maintaining these fish. The results presented by Jenkins and his colleagues show just how much can be achieved with careful analysis and control.

Three female and five male red drum adults were held in an indoor tank and subjected to a standard, 6-month photothermal conditioning cycle. Once the appropriate temperature and photoperiod were reached (25°C, 13 h light), the fish began to spawn in the tank of their own volition. At this point, temperature and photoperiod were held constant from day to day for approximately 3 months. During this 3-month period, a total of 73 spawns occurred within the tank. Fish spawned at will, with daily egg production ranging from nothing to as much as 1.7 million eggs. Total production during this time period was estimated at 47 million eggs.

By the end of the 3-month period, however, spawning patterns had become erratic and unpredictable. At that time, conventional wisdom in most red drum hatcheries and research laboratories would have suggested the conditioning cycle should be re-initiated, based on the reduced frequency of spawning. The authors, however, tried something different. They maintained a 13-h day length, but initiated a repeating 7-day thermal regime: 4 days at 25°C followed by 3 days at 21°C. The red drum would spawn within one day each time the temperature was raised and continue to spawn until the temperature was lowered again.

This 7-day temperature regime was used to spawn the fish on a weekly schedule for 14 months. A total of 87 million eggs were produced in 154 spawns. Egg viability was estimated at 74.7%, better than that obtained during the initial 3-month spawning period (68.4%). This type of extended production period is possible because red drum are able to ovulate buoyant, mature eggs (Figs 9.22 and 9.23) and then immediately prepare more oocytes for ovulation, to accommodate their natural pattern of spawning several or more times during their reproductive season (mid-to-late fall).

9.4.3 A case study: the striped bass (*Morone saxatilis*) and its hybrids

In contrast, controlled reproduction is still a major constraint for another fish widely cultured in the US and elsewhere: hybrid striped bass (*Morone saxatilis* [striped bass] × *M. chrysops* [white bass]). These hybrids were first produced through strip spawning hormone-injected broodfish based on the pioneering work of Stevens (1966) and Bayless (1972). Bishop (1975) reported on the successful reproduction of hormone-injected *M. saxatilis* broodfish held in circular tanks. Although

Fig. 9.22 A simple egg collection device at the outflow of a tank used for photothermal conditioning and spawning. When lowered, the plastic channel diverts outflow into a screened collection chamber, which in turn drains into the central drain line for a number of conditioning/spawning tanks. (Photo by the author.)

this method reduced broodfish handling mortality, it was still dependent on the capture and removal of wild fish from spawning stocks. Accordingly, in 1983 the federal Joint Subcommittee on Aquaculture cited "the non-availability of seedstock" as "the major constraint to private striped bass aquaculture in the United States."

To date, the industry still relies almost entirely on wild fish. Estimates of the current industry demand for hybrid fry in the US range as high as 40–50 million annually, based on expected survival and additional sales to other countries. In spite of continued reliance on wild broodstock, efforts by a number of institutions have been ongoing for several decades to refine techniques for induced spawning in striped bass, related species and their hybrids.

Smith and Jenkins (1984) reported on controlled spawning of F_1 hybrid bass through the use of photothermal conditioning and injections of hCG. Similarly, Henderson-Arzapalo and Colura (1987) utilized photothermal conditioning and subsequent hCG injection to induce maturation and spawning in striped bass. In recent years, however, researchers working with striped bass (as in many other species) have found better ways than hypodermic injection of hormones to force the completion of internal reproductive processes leading to spawning.

Hodson and Sullivan (1993) described the maturation and spawning of domestic and wild striped bass using a combination of implants and injections. They used cholesterol–cellulose implants, containing GnRHa, in females that had been raised in captivity and in less mature

Fig. 9.23 In order to evaluate specific strains, lines or families, or to produce novel genotypes such as gynogens or polyploids, it is not only necessary to obtain suitable gametes from the species in question but also to successfully incubate the resultant embryos. Shown here is a "MacDonald"-type hatching jar, which provides a continuous up-welling current for incubation of non-buoyant eggs. (Photo by the author.)

wild-caught females. Each fish was implanted with one fast-release (80% cholesterol) and one slow-release (95% cholesterol) pellet, and injected with hCG after 1–3 days when final oocyte maturation began. The resulting egg and fry production was comparable to, if somewhat more expensive than, results expected from fully mature wild-caught females.

Woods *et al.* (1995) reported on the use of implants containing a GnRH analog in captive white bass males and laboratory-reared striped bass females. The implants induced five of eight female stripers in an advanced state of maturity to spawn in tanks with white bass males, which had been injected with hCG in addition to receiving implants. Only two of the spawning females actually produced fertilized eggs resulting in hybrid offspring, apparently as a result of a high incidence of over-ripening.

At times, what may seem like "more spawning" in captive broodstocks doesn't necessarily translate into more reproduction. A fine illustration of this principle, utilizing *M. saxatilis*, was presented by

Blythe et al. (1994). These researchers evaluated three different photothermal conditioning cycles on captive-reared striped bass: a 12-month "reference" cycle, with photoperiod and water temperature controlled to mimic natural conditions for the mid-Atlantic region of eastern North America, and two compressed variations of the reference cycle, reduced to either 9 months or 6 months.

The broodstock used in the study were hatchery-produced striped bass from the National Fisheries Research Center in Leetown, West Virginia. They had been reared for 6 years in raceways from an average size of approximately 13 cm. Brood fish were approximately 6 years old and sexually mature when the evaluations began, weighing 2.5 kg on average. Five environmental chambers were used for the study, each housing a 1.9 m diameter tank with a recirculating system to maintain water quality. Four males and four females were stocked into each tank. Two chambers were assigned to each of the compressed cycles and the fifth chamber was operated on the 12-month reference cycle. Photoperiod and water temperature were controlled for each chamber/tank by a series of microcomputers.

Each month, sonograms were performed on each fish and ultrasound images were recorded on videotape. Fish were also squeezed (males) or catheterized (females) to evaluate their state of maturation. In each photothermal regime, once the natural spawning temperature and photoperiod were reached fish were held under constant conditions until females reached a stage of maturation suitable for hormone injection to stimulate ovulation. This maturation holding period required the extension of the spawning cycles to 7.4, 9.8, and 12.7 months for the 6-, 9-, and 12-month treatments, respectively.

Once maturation had progressed to a suitable point, females were transferred to a spawning tank with spermiating males and all fish were injected with hCG after one week. Fish were either strip spawned (upon ovulation) or allowed to spawn in the tank of their own volition. Egg diameter, ovarian diameter, testicular dimensions, and spawning characteristics such as fertility, fecundity and latent time were all compared among fish subjected to the different conditioning cycles.

In this situation, more spawning cycles resulted in less reproduction. All males in the 9- and 12-month cycles matured and spermiated, while only 88% of those in the 6-month cycle did so. While all females subjected to the 9- and 12-month cycles matured (n = 8 and 4, respectively), only four of the eight females in the 6-month cycle tanks did so. While seven of eight females on the 9-month cycle and all four females on the 12-month cycle ovulated, only three of the four mature females on the 6-month cycle ovulated.

Egg size, ovarian diameters and testicular diameters were all related to length of the conditioning cycles. Maximum ovarian diameters were 19.6 mm, 27.4 mm and 24.5 mm for females subjected to the 6-, 9-, and 12-month cycles, respectively. Corresponding maximum egg diameters were 721 µm, 945 µm, and 1073 µm, respectively. Testicular diameters of males (16.2 mm, 23.6 mm and 19.9 mm for the three cycles, respectively) followed a pattern similar to that of ovarian diameters.

The authors also examined what would happen on a subsequent 6-month cycle immediately following the initial 6-month treatment. "Less" quickly became "lesser." Of eight females, five matured, but only two of the five ovulated: a female that had not spawned during the first 6-month cycle and a replacement fish for a female that had died during the first cycle. Maximum egg and ovarian diameters, 794 µm and 22.8 mm, were similar to those recorded in the initial 6-month cycle. Only 43% of the males spermiated when subjected to the second 6-month cycle. Additionally, testicular diameters during the second cycle were 12.8 mm, compared to 16.2 mm during the initial period.

Fecundity and fertility of those fish that did spawn were similar for both spawning methods and all conditioning cycles, including the second 6-month cycle. The lesson here? Under the conditions used in this study, compressing the annual photothermal conditioning regime for striped bass below a 9-month cycle was almost certainly counter-productive, at least in terms of numbers of eggs or fry produced on a per-month basis. When evaluating photothermal regimes for induced spawning in an undocumented species, a full 12-month cycle should be evaluated and refined initially prior to subsequent efforts to efficiently compress the cycle. As an example, if year-round striped bass egg and fry production were required, the maximum return in numbers of eggs and fry might best be attained by staggering 9- or 12-month cycles in such a way as to always have some tanks of broodfish spawning or about to spawn.

9.4.4 Induction with asynchronous ovarian development

In contrast to red drum, which are able to ovulate mature eggs and then immediately prepare more oocytes for ovulation, fish like striped bass generally only spawn once per year, at some brief point during a very limited spawning season. This results in a hit-or-miss success rate, but also a simplified technical approach to spawning induction. Many other aquacultured species with reproductive patterns similar to those of red drum, like the gilthead seabream (*Sparus aurata*), also ovulate some portion of their oocytes and subsequently ovulate more, newly maturing oocytes within one to several days. Mylonas *et al.* (1995) demonstrated a similar pattern of asynchronous ovarian development in the American shad (*Alosa sapidissima*), using wild-caught broodstock and hormone-releasing implants.

In the Jenkins *et al.* (1997) study, environmental conditions of temperature and photoperiod alone were sufficient to induce spawning in adult red drum. Similar conditions, with a little help from a single hormonal injection, were also sufficient to induce spawning in the striped bass Blythe *et al.* (1994) were working with. The shad broodstock Mylonas *et al.* (1995) were working with were wild-caught, however, and once they were pulled from their spawning grounds the internal metabolic pathways that would have led to spawning were broken down due to capture and handling stress and could only be patched together using hormonal administration. This is often the case when wild broodstock must be captured during or immediately prior to their spawning seasons.

In order to take advantage of the shad's asynchronous ovarian development and promote multiple ovulations over an extended period while avoiding the need for repeated handling and hormone injections, Mylonas *et al.* (1995) used extended-release implants to deliver the needed stimulus over time for repeated ovulations. The resulting reduction in day-to-day handling and capture stress resulted in much better shad survival and reproductive success over the spawning period. Another example of how less work can result in more egg and fry production, with the proper tools, knowledge and planning.

9.4.5 Hormone injection of various fish

Groupers
Watanabe *et al.* (1995) reported on controlled breeding of the Nassau grouper (*Epinephelus striatus*) using injections of hCG, LHRHa, and carp pituitary homogenate (CPH) separately or in various combinations. Two-injection protocols, based on an initial (priming) dose and a follow-up (resolving) dose were evaluated. Injection treatments were hCG alone, LHRHa alone, both compounds in combination (each as priming or resolving dose) and CPH in combination with LHRHa. All treatments produced successful spawnings with fertilization rates of 50% or greater. In a follow-up study, Head *et al.* (1996) reported successful spawning induction in female *Epinephelus striatus* through the administration of a priming dose of 1000 IU hCG kg^{-1} body weight followed 24 h later by a resolving dose of 500 IU hCG kg^{-1} body weight. Estimated numbers of eggs spawned ranged from 200 000 to 2 000 000 per female, with fertilization and hatching rates of 17.9–80.6% and 68.3–90.1%, respectively. While the authors correctly pointed out that this level of fecundity would minimize broodstock requirements for commercial production, the potential for inbreeding could become quite high if captive stocks were ultimately selected for induced spawning.

Catfish
Alok *et al.* (1995) compared the efficacy of single intramuscular injections of pituitary extract and sGnRHa, both with and without supplemental domperidone, in the catfish *Clarias batrachus*. Although 50 µg of sGnRHa kg^{-1} body weight resulted in ovulation when combined with 5 mg of domperidone kg^{-1} body weight, injections of sGnRHa alone failed to induce maturation and ovulation, even at doses of 500 µg kg^{-1} body weight. Pituitary extract injection at a dose of 40 mg kg^{-1} body weight produced results comparable to those obtained from the sGnRHa + domperidone combination. No compound, dose or combination resulted in spawning, and ovulated eggs were stripped for manual fertilization.

Kelly and Kohler (1996) documented the utility of hormonal administration in addition to photothermal manipulation in channel catfish

Ictalurus punctatus. Under a variety of photothermal regimes, half of the broodstock pairings designated for spawning were subjected to injections of LHRHa and hCG while half were allowed to spawn naturally. Hormone-induced females laid fewer eggs than non-injected females, either in-season or out-of-season, suggesting the photothermal conditioning may have been sufficient to induce maturation and spawning and that stress from capture and handling associated with hormonal injections may have actually reduced spawning success.

The Asian catfish *Clarias macrocephalus* normally spawns during a period ranging mainly from July through September. Tan-Fermin, Pagador & Chavez (1997) examined the effects of hormonal administration before, during, after and between spawning seasons on induced spawning in this species. At each point in time, they evaluated the effects of intramuscular injections of LHRHa at a rate of 0.05 $\mu g\ g^{-1}$ body weight, alone or in conjunction with pimozide (at 1 $\mu g\ g^{-1}$ body weight). In all seasonal periods, only those fish injected with a combination of LHRHa and pimozide spawned. Initial egg size, ovulation rates, egg production, and fertilization, hatching and survival rates all indicated better chances of successful fry production before or during the spawning season than between spawning seasons.

Bates and Tiersch (1998) reported on the use of LHRHa in inducing ovulation and spawning in *I. punctatus*. After a large number of females were injected, half were paired with males for tank spawning and half were grouped in tanks without males present. Spawning (or egg-laying) success was 41% for the paired fish and 58% within the all-female groups. Elapsed time between injection and spawning (also known as "latency") was similar for the two groups. As eggs were taken from ovulating females in the all-female groups and fertilized artificially, resulting fertilization rates were $43 \pm 20\%$ for paired fish and $16 \pm 37\%$ for grouped females.

Working with the European catfish, *Silurus glanis*, Brzuska and Adamek (1999) used single and multiple (two) injections of carp pituitary extract and LHRHa with pimozide (a dopamine receptor antagonist) to stimulate ovulation. While all fish treated with LHRHa and pimozide ovulated, only 60–67% of those treated with carp pituitary did so. Silverstein *et al.* (1999) also examined the effects of combining or excluding pimozide with LHRHa injections in the channel catfish. Their results indicated the number of spawns was significantly higher in fish that were injected with pimozide and LHRHa than in those injected only with LHRHa (83% compared to 44%, respectively). Percent hatch did not differ significantly between these treatments.

Perch and their relatives

Kucharczyk *et al.* (1996) reported on the use of hCG with or without carp pituitary to induce spawning in the European perch *Perca fluviatilis*. Best results were obtained when these compounds were administered in combination, with dosages of 4 mg carp pituitary and 500–700 IU hCG kg^{-1} for females and half those amounts for males, divided among two to three injections. A number of hormonal and dopamine blocker combinations were subsequently evaluated by these authors for their effect on ovulation and spawning in *Perca fluviatilis* (Kucharczyk et al. 1998). The most effective combination, in terms of spawning success, was 75 IU kg^{-1} body weight of follicle stimulating hormone and luteinizing hormone in conjunction with 10 mg kg^{-1} body weight of metoclopramide.

Malison et al. (1998) captured wild *Stizostedion vitreum* in the autumn and transferred these fish into earthen ponds for overwintering. Approximately 10, 6 and 3 weeks prior to the normal spawning season in April, fish were moved into indoor tanks and acclimated gradually from 2–10°C, while photoperiod was held at 12 h light: 12 h darkness. Females were injected with either LHRHa, hCG, or 17,20-P. At 10 weeks prior to the spawning season, results were mixed: hCG resulted in the highest number of ovulating females (three out of five fish). Subsequently, at 6 and 3 weeks prior to the beginning of the natural spawning season all but one injected fish spawned. Fish injected with LHRHa and hCG generally produced eggs of acceptable quality, while those of fish injected with 17,20-P were generally small and exhibited poor survival.

Various other species

While releasing hormones and their analogs often provide results where pituitary extracts are ineffective, the opposite can also be true depending on the species and circumstances in question. Kurup et al. (1993), working with *Labeo dussumieri*, indicated that a priming dose of 4 mg kg^{-1} and a resolving dose of 8 mg kg^{-1} of carp pituitary extract in females and a single carp pituitary extract injection of 4 mg kg^{-1} in males resulted in successful spawning induction, while hCG or LHRH failed to induce ovulation. Further trials established that a priming dose of carp pituitary extract in conjunction with pimozide, followed by hCG or LHRH injections, resulted in successful induction, although fertilization and hatching rates were lower than in spawns induced with carp pituitary extract.

Godinho et al. (1993) evaluated the use of hCG alone and in combination with mullet pituitary extract to induce spawning in the mullet *Mugil platanus* in Brazil. They reported that a two-injection dose of hCG (20 and 40 IU g^{-1} body weight at 24 h intervals), as well as doses of 10 or 20 mg of mullet pituitary extract kg^{-1} body weight in combination with 30 IU hCG g^{-1} body weight all resulted in high rates of spawning, fertilization and hatching.

Glubokov et al. (1994) examined the use of neuroleptics, specifically pimozide, sulpiride, metoclopramide and isofloxythepin in conjunction with releasing hormones (LHRHa and GnRH-A) on spawning induction in the Pacific mullet, *Mugil so-iuy*. They determined the most efficient approach for induction in this species, with these particular compounds, involved doses of 1–8 mg kg^{-1} body weight of neuroleptic compounds, followed 4 and 28 h later by injections of 25 and 50 μg of GnRH-A kg^{-1} body weight, respectively, in mature females.

Although the red drum *Sciaenops ocellatus* may be the most widely studied of the sciaenids, many species in this group may have potential for aquaculture. Battaglene and Talbot (1994) reported on induced spawning and larval rearing in the Australian sciaenid *Argyrosomus hololepidotus*, commonly known as the mulloway. Pond-held mulloway were administered a single injection of hCG, at 1000 IU kg^{-1} body weight for females and a quarter of that dose for males. Eggs and milt were stripped following ovulation, with females producing between 0.9 and 1.0 million eggs apiece. As in other species, such high levels of fecundity result in low broodstock requirements, but also potentially low genetic diversity among resultant production stocks.

Although producers and researchers generally have no trouble encouraging the various species of tilapia to spawn under a variety of conditions, spawning generally occurs on a more or less asynchronous basis and there are occasions when more synchronized spawning would be desirable to allow for specific research needs or reduction of labor in egg and fry collection. Roderick *et al.* (1996) reported on the use of a number of ovulatory inducers to enable more reliable egg collection in *Oreochromis niloticus*. While LHRHa (at 30 mg 100 g^{-1} body weight) and fertirelin acetate (at 1–22.5 mg 100 g^{-1} body weight) injections were effective at increasing the number of spawns over a period of 8 days, spawning patterns were still somewhat asynchronous.

As has been seen in other species, Alok *et al.* (1997) established the requirement for supplemental addition of a dopamine blocker (domperidone) when administering a releasing hormone analog (in this case sGnRH-a) to common carp (*Cyprinus carpio*). When working with mature fish at the end of the normal spawning season, injections of releasing hormone analog or domperidone alone resulted in no spawns, but the two compounds in combination at rates of 10 and 20 µg kg^{-1} body weight respectively were effective in inducing ovulation and spawning.

Culture of various suckers has become an important segment of bait- and foodfish industries in several regions. Ludwig (1997) reported on the use of hCG injections to induce ovulation and spawning in the white sucker (*Catostomus commersoni*) and the spotted sucker (*Minytrema melanops*). In both species, injections of 1000 IU of hCG kg^{-1} of body weight daily for 3 to 5 days resulted in successful spawning.

Implant applications

For many species, results can be considered acceptable even if the fish do not actually spawn of their own volition. Successful ovulation is often sufficient to allow for production of large numbers of fry. Berlinsky *et al.* (1996) reported on induced ovulation in the southern flounder (*Paralichthys lethostigma*), through the use of GnRHa implants. Of 12 females with mature ovarian follicles (>500 µm diameter), 11 ovulated within 90 h after implantation. Eggs, however, had to be manually stripped and fertilized artificially. Although fertility varied, the methods described represented an important step forward in the commercial development of this promising flatfish.

Hassin et al. (1997) encountered similar limitations in their work with the white grouper (*Epinephelus aeneus*). Although captive fish grew rapidly and adjusted to artificial diets, final oocyte maturation, ovulation and spawning did not occur until sustained release implants of GnRHa were used. Even then, although ovulation was induced, natural spawning did not take place, requiring manual stripping and artificial fertilization of ovulated eggs. In these situations, careful observation (under conditions of high visibility) and frequent monitoring (with ease of capture and handling) are required to detect ovulation in time to avoid over-ripening of eggs.

In trials with a cold-water flatfish, Larsson et al. (1997) reported impressive results from implanting yellowtail flounder (*Pleuronectes ferrugineus*) with GnRHa. Intramuscular implants containing 75–224 µg kg^{-1} female body weight, either as cholesterol pellets or biodegradable microspheres, resulted in earlier ovulation and multiple batches of high-quality eggs. Females were more closely synchronized, with shorter interovulatory periods, and egg production rate was approximately doubled. The authors noted that, as would be anticipated based on the physiological pathway model, plasma levels of the hormone estradiol-17 beta were elevated 96 h after implantation.

9.4.6 Crustaceans

Aiken (1969) reported on findings from a well-designed study examining the effects of photoperiod and temperature (separately and in combination) on maturation and oviposition in female *Orconectes virilis*. A combination of cold (4°C) water and total darkness for 4 months or longer was sufficient to allow ovarian maturation and subsequent oviposition at 20°C and a 17L:7D photoperiod. However, when applied separately, each environmental condition (total darkness at 20°C, or 4°C water with an 18L:6D photoperiod) resulted in elevated levels of mortality and failed to produce acceptable levels of maturation and spawning. Once ovarian maturation had been attained by holding for 4 months or longer in total darkness at 4°C water, temperature rather than photoperiod was shown to be the stimulus required for oviposition.

McRae and Mitchell (1997) examined methods to encourage ovarian rematuration following spawning in the Australian crayfish *Cherax destructor*. Their findings suggested that rematuration could not proceed until eggs or neonates from the previous spawn had been released. Additionally, confinement in isolation under unchanging environmental conditions also prevented rematuration. A drop in temperature or the physical presence of males both resulted in stimulation of ovarian development. Injection of serotonin did not result in ovary stimulation, and rematuration required hepatosomatic indices of 7% or greater.

Unilateral eyestalk ablation has long been recognized as a practical means to induce spawning in female penaeid shrimp. Perhaps the most physically expedient method involves the enucleation of the eyestalk through the distal end of the eye (Treece & Fox 1993). An important point to consider is the phase in the molt cycle during which ablation is to take place. During the pre-molt phase, ablation normally results in acceleration of molting. During the immediate post-molt period, ablation can result in mortality. Ablation during the intermolt period, however, generally leads to maturation and subsequent spawning if high standards of water quality, proper temperature and appropriate light conditions are maintained. While ablation of male penaeid shrimp is generally not necessary, it has been associated with precocious maturation and increased mating frequency in immature males of several species (Alikunhi et al. 1975; Chamberlain & Lawrence 1981).

Vaca and Alfaro (2000) compared the injection of serotonin with unilateral eyestalk ablation as means of inducing ovarian maturation and spawning in wild *Penaeus vannamei*. Although injection doses of both 15 and 50 µg g^{-1} body weight induced maturation and spawning, the ablation protocol resulted in more rapid and reliable responses.

9.4.7 Mollusks

Although thermal shocks and stimulation by serotonin injection into the gonads are both widely used techniques to induce spawning in bivalve mollusks, simpler means have come to light for some species in recent years. Desrosiers and Dube (1993) reported on the use of flowing sea water to induce both male and female giant scallops (*Placopecten magellanicus*) to shed their gametes. While chemical stimulation by serotonin injection resulted in 100% induction of gamete release by male scallops, flowing sea water was 93% effective for males and 100% effective for females. Additionally, the total number of spermatozoa released under mechanical (flowing sea water) stimulation was roughly one order of magnitude higher than the production from serotonin-injected scallops.

Madrones-Ladja (1997) compared the utility of intragonadal injections of serotonin with addition of UV-irradiated sea water as means to induce spawning in the window-pane shell *Placuna placenta*. Although serotonin injection produced high numbers of spawned eggs within 15–30 min, UV-irradiated sea water resulted in only slightly lower numbers of eggs over a 1-h period. Madrones-Ladja concluded that the use of UV-irradiated sea water to induce spawning was simpler to perform and could be applied more easily to mass-spawning experiments.

Ellis (1998) provided an excellent review of methods for spawning induction in the tridacnid giant clams. Since spawning in these species tends to follow diel and lunar cycles, being highest in the afternoons during full and new moons, induction is generally undertaken to coincide with these periods. The most practical method of inducing giant clams to spawn involves heat stressing mature animals. Clams can be held in shallow water or even out of the water under a hot sun for 1–2 h and subsequently returned to normal conditions for 30–60 min. This heat stress may be repeated as many as four times in one day, usually in the morning. As long as internal temperatures do not approach 35°C, giant clams generally tolerate this practice and exhibit no long-term effects.

Alternative approaches to inducing giant clams to spawn involve injecting serotonin into the gonads of mature animals or adding an extract of gonadal material to holding tanks (Ellis 1998). A solution of 15.7 µg of serotonin is dissolved in filtered (1 µm) sea water and usually injected using a 150 µm needle. Alternately, approximately 20 g of gonadal material from a sacrificed clam is blended with filtered sea water and filtered (200 µm), then added directly to shallow, static holding tanks or introduced directly into the incurrent siphon or each clam using a syringe (again under shallow, static conditions).

9.5 REFERENCES

Aiken, D.E. (1969) Ovarian maturation and egg laying in the crayfish *Orconectes virilis*: influence of temperature and photoperiod. *Canadian Journal of Zoology,* **47**, 931–935.

Alikunhi, K.H., Poernomo, A., Adisufresno, S., Budiono, M. & Busman, S. (1975) Preliminary observations on induction of maturity and spawning in *Penaeus monodon* (Fabricius) and *Penaeus merguiensis* (deMan) by eyestalk ablation. *Bulletin of the Shrimp Culture Research Center, Jepara,* **1** (Suppl. 1), 1–11.

Alok, D., Pillai, D. & Garg, L.C. (1997) Effect of D-Lys^6 salmon sGnRH alone and in combination with domperidone on the spawning of common carp during the late spawning season. *Aquaculture International,* **5** (Suppl. 4), 369–374.

Alok, D., Pillai, D., Talwar, G.P. & Garg, L.C. (1995) D-Lys^6 salmon gonadotropin-releasing hormone analogue-domperidone induced ovulation in *Clarias batrachus* (L.) *Asian Fisheries Science,* **8** (Suppl. 3 & 4), 263–266.

Almendras, J.M., Duenas, C., Nacario, J., Sherwood, N., & Crim, L.W. (1988) Sustained hormone release. III. Use of gonadotropin releasing hormone analogues to induce multiple spawnings in sea bass, *Lates calcarifer. Aquaculture,* **74**, 97–111.

Arnold, C.R., Bailey, W.H., Williams, T.D., Johnson, A. & Laswell, J.L. (1976) Laboratory spawning and larval rearing of red drum and southern flounder. *Proceedings of the Annual Conference of Southeastern Fish and Wildlife Agencies,* **31**, 437–440.

Barry, T.P., Malison, J.A., Lapp, A.F. & Procarione, L.S. (1995) Effects of selected hormones and male cohorts on final oocyte maturation, ovulation, and steroid profiles in walleye (*Stizostedion vitreum*). *Aquaculture,* **138** (Suppl. 1–4), 331–347.

Bates, M.C. & Tiersch, T.R. (1998) Preliminary studies of artificial spawning of channel catfish as male-female pairs or all-female groups in recirculating systems. *Journal of the World Aquaculture Society,* **29** (Suppl. 3), 325–334.

Battaglene, S.C. & Selosse, P.M. (1996) Hormone-induced ovulation and spawning of captive and wild broodfish of the catadromous Australian bass, *Macquaria novemaculeata* (Steindachner), (Percichthyidae). *Aquaculture Research,* **27** (Suppl. 3), 191–204.

Battaglene, S.C. & Talbot, R.B. (1994) Hormone induction and larval rearing of mulloway, *Argyrosomus hololepidotus* (Pisces:Sciaenidae). *Aquaculture,* **126** (Suppl. 1–2), 73–81.

Bayless, J.D. (1972) *Artificial Propagation and Hybridization of Striped Bass,* Morone saxatilis *(Walbaum).* South Carolina Wildlife Resources Department.

Berlinsky, D.L., King, W.V., Smith, T.I.J., Hamilton, R.D., Holloway, J. & Sullivan, C.R. (1996) Induced ovulation of southern flounder (*Paralichthys lethostigma*) using gonadotropin releasing hormone analogue implants. *Journal of the World Aquaculture Society*, **27** (Suppl. 2), 143–152.

Bishop, E.D. (1975) The use of circular tanks for spawning striped bass (*Morone saxatilis*). *Proceedings of the Annual Conference of the Southeastern Association of Fish and Wildlife Agencies*, **28**, 35–44.

Blythe, W.G., Helfrich, L.A. & Libey, G. (1994) Induced maturation of striped bass *Morone saxatilis* exposed to 6, 9, and 12 months photothermal regimes. *Journal of the World Aquaculture Society*, **25** (Suppl. 2), 183–192.

Brzuska, E. & Adamek, J. (1999) Artificial spawning of European catfish, *Silurus glanis* L.: stimulation of ovulation using LHRH-a, Ovaprim and carp pituitary extract. *Aquaculture Research*, **30** (Suppl. 1), 59–64.

Cantin, M.-C. (1986) Oocyte development and induced spawning of smallmouth bass (*Micropterus dolomieui*) in response to photoperiod and temperature manipulation. Preliminary results. In: *Book of Abstracts: Aquaculture '86 – Reno*, World Aquaculture Society, Baton Rouge.

Carrillo, M., Zanuy, S., Prat F. et al. (1995) Sea bass (*Dicentrarchus labrax*). In: *Broodstock Management and Egg and Larval Quality* (eds N.R. Bromage & R.J. Roberts), pp. 138–168. Blackwell Science Ltd, Oxford, UK.

Chamberlain, G.W. & Lawrence, A.L. (1981) Maturation, reproduction, and growth of *Penaeus vannamei* and *P. stylirostris* fed natural diets. *Journal of the World Mariculture Society*, **12** (Suppl. 1), 209–224.

Clemens, H.P. & Sneed, K.E. (1957) *Spawning behavior of the channel catfish Ictalurus punctatus*. U.S. Fish and Wildlife Service Special Scientific Report – Fisheries, no. 219. 11 pp. Department of the Interior, Washington D.C.

Copeland, P.A. & Thomas, P. (1989) Control of gonadotropin release in the Atlantic croaker (*Micropogonias undulatus*): evidence for lack of dopaminergic inhibition. *General and Comparative Endocrinology*, **74**, 479–483.

Crim, L.W. & Glebe, B.D. (1984) Advancement and synchrony of ovulation in Atlantic salmon with pelleted LHRH analog. *Aquaculture*, **43**, 47–56.

Crim, L.W., Sutterlin, A.M., Evans, D.M. & Weil, C. (1983) Accelerated ovulation by pelleted LHRH analogue treatment of spring-spawning rainbow trout *Salmo gairdneri* held at low temperature. *Aquaculture*, **35**, 299–307.

Desrosiers, R.R. & Dube, F. (1993) Flowing sea water as an inducer of spawning in the sea scallop, *Placopecten magellanicus*, (Gmelin, 1791). *Journal of Shellfish Research*, **12** (Suppl. 2), 263–265.

Ellis, S. (1998) *Spawning and Early Larval Rearing of Giant Clams (Bivalvia: Tridacnidae)*. CTSA Publication no. 130. Center for Tropical and Subtropical Aquaculture, Waimanalo, Hawaii.

Fagbenro, O.A., Salami, A.A. & Sydenham, D.H.J. (1992) Induced ovulation and spawning in the catfish, *Clarias isheriensis*, using pituitary extracts from nonpiscine sources. *Journal of Applied Aquaculture*, **1** (Suppl. 4), 15–20.

Falls, W.W., Dennis, C.W., Hindle, P.A. & Young, J.C. (1997) Refinements in methods of inducing spawning in cultured red drum, *Sciaenops ocellatus* (L.), in Florida. In: *Book of Abstracts: World Aquaculture '97 – Seattle*, World Aquaculture Society, Baton Rouge.

Glubokov, A.I., Kouril, J., Mikodina, E.V. & Barth, T. (1994) Effect of synthetic GnRH analogues and dopamine antagonists on the maturation of Pacific mullet, *Mugil so-iuy* Bas. *Aquaculture and Fisheries Management*, **25** (Suppl. 4), 419–425.

Godinho, H.M., Kavamoto, E.T., Andrade-Talmelli, E.F. & Serralheiro, P.C.S. (1993) Induced spawning of the mullet *Mugil platanus* Guenther, 1880, in Cananeia, Sao Paulo, Brazil. *Boletim do Instituto de Pesca*, **20**, 59–66.

Harvey, B.J. & Kelley, R.N. (1988) Practical methods for chilled and frozen storage of tilapia spermatozoa. In: *The Second International Symposium on Tilapia in Aquaculture* (eds R.S.V. Pullin, T. Bhukaswan, K. Tonguthai & J.L. Maclean), pp. 187–189. ICLARM – International Center for Living Aquatic Resources Management, Manila, Philippines.

Hassin, S., de Monbrison, D., Hanin, Y., Elizur, A., Zohar, Y. & Popper, D.M. (1997) Domestication of the white grouper (*Epinephelus aeneus*) 1. Growth and reproduction. *Aquaculture*, **156** (Suppl. 3–4), 309–320.

Head, W.D., Watanabe, W.O., Ellis, S.C. & Ellis, E.P. (1996) Hormone induced multiple spawning of captive Nassau grouper broodstock. *The Progressive Fish-Culturist*, **58** (Suppl. 1), 65–69.

Henderson-Arzapalo, A. & Colura, R.L. (1987) Laboratory maturation and induced spawning of striped bass. *The Progressive Fish-Culturist*, **49**, 60–63.

Hodson, R.G. & Sullivan, C.V. (1993). Induced maturation and spawning of domestic and wild striped bass, *Morone saxatilis* (Walbaum), broodstock with implanted GnRH analogue and injected with hCG. *Aquaculture and Fisheries Management*, **24** (Suppl. 3), 389–398.

Hoff, F., Rowell, C. & Pulver, T. (1972) Artificially induced spawning of the Florida Pompano under controlled conditions. *Proceedings of the World Mariculture Society*, **3**, 53–64.

Jenkins, W.E., Bridgham, C.B. & Smith, T.I.J. (1997) Closely controlled spawning of captive red drum (*Sciaenops ocellatus*) brood stock. In: *Book of Abstracts: World Aquaculture '97 – Seattle*. World Aquaculture Society, Baton Rouge.

Joint Subcommittee on Aquaculture (1983) Striped bass species plan. In: *National Aquaculture Development Plan. Volume II*, pp. 136–158. Federal Coordinating Council on Science, Engineering and Technology, Washington, D.C.

Kelly, A.M. & Kohler, C.C. (1996) Manipulation of spawning cycles of channel catfish in indoor water-recirculating systems. *The Progressive Fish-Culturist*, **58** (Suppl. 4), 221–228.

Kobayashi, M. & Stacey, N. (1993) Prostaglandin-induced female spawning behavior in goldfish (*Carassius auratus*) appears independent of ovarian influence. *Hormones and Behavior*, **27**, 38–55.

Kucharczyk, D., Kujawa, R., Luczynski, M., Glogowski, J. & Babiak, I. (1997) Induced spawning in bream, *Abramis brama* (L.), using carp and bream pituitary extract and hCG. *Aquaculture Research*, **28** (Suppl. 2), 139–144.

Kucharczyk, D., Kujawa, R., Mamcarz, A., Skrzypczak, A. & Wyszomirska, E. (1996) Induced spawning in perch, *Perca fluviatilis* L., using carp pituitary extract and HCG. *Aquaculture Research*, **27** (Suppl. 11), 847–852.

Kucharczyk, D., Kujawa, R., Mamcarz, A., Skrzypczak, A. & Wyszomirska, E. (1998) Induced spawning in perch, *Perca fluviatilis* L., using FSH + LH with pimozide or metoclopramide. *Aquaculture Research*, **29** (Suppl. 2), 131–136.

Kurup, B.M., Nair, C.M. & Kuriakose, B. (1993) Preliminary observations on induced breeding of *Labeo dussumieri* (Val.) using carp pituitary extract and other hormones. *Journal of Aquaculture in the Tropics*, **8** (Suppl. 2), 255–262.

Larsson, D.G.J., Mylonas, C.C., Zohar, Y. & Crim, L.W. (1997) Gonadotropin-releasing hormone analogue (GnRH-A) induces multiple ovulations of high-quality eggs in a cold-water, batch-spawning teleost, the yellowtail flounder (*Pleuronectes ferrugineus*). *Canadian Journal of Fisheries and Aquatic Sciences/Journal Canadien des Sciences Halieutiques et Aquatiques*, **54** (Suppl. 9), 1957–1964.

Lasker, R. (1974) Induced maturation and spawning of marine fish at the southwest fisheries center, La Jolla, California. *Proceedings of the World Mariculture Society*, **5**, 313–318.

Lawson, T.B., Drapcho, C.M., McNamara, S., Braud, H.J. & Wolters, W.R. (1989) A heat exchange system for spawning red drum. *Aquacultural Engineering*, **8** (Suppl. 3), 177–208.

Lee, C.S., Tamaru, C.S., Kelley, C.D. & Banno, J.E. (1986a) Induced spawning of milkfish, *Chanos chanos*, by a single application of LHRH-analogue. *Aquaculture*, **58**, 87–98.

Lee, C.S., Weber, G. & Tamaru, C.S. (1986b) Studies on the maturation and spawning of milkfish, *Chanos chanos* in a photoperiod-controlled room. *Book of Abstracts: Aquaculture '86 – Reno*, World Aquaculture Society, Baton Rouge.

Leu, M.-Y. & Chou, Y.-H. (1996) Induced spawning and larval rearing of captive yellowfin porgy, *Acanthopagrus latus* (Houttuyn). *Aquaculture*, **143** (Suppl. 2), 155–166.

Ludwig, G.M. (1997) Induced spawning in captive white sucker, *Catostomus commersoni*, and spotted sucker, *Minytrema melanops*. *Journal of Applied Aquaculture*, **7** (Suppl. 3), 7–17.

Madrones-Ladja, J.A. (1997) Notes on the induced spawning, embryonic and larval development of the window-pane shell, *Placuna placenta* (Linnaeus, 1758), in the laboratory. *Aquaculture*, **157** (Suppl. 1–2), 135–144.

Malison, J.A., Procarione, L.S., Kayes, T.B., Hansen, J.F. & Held, J.A. (1998) Induction of out-of-season spawning in walleye (*Stizostedion vitreum*). *Aquaculture*, **163** (Suppl. 1–2), 151–161.

McCarty, C.E. (1990) Design and operation of a photoperiod/temperature spawning system for red drum. In: *Red Drum Aquaculture* (eds G.W. Chamberlain, R.J. Miget & M.G. Haby), pp. 44–45. Texas A&M University.

McRae, T.G. & Mitchell, B.D. (1997) Control of ovarian rematuration in the yabby, *Cherax destructor* (Clark). *Freshwater Crayfish*, **11**, 299–310.

Muiruri, R.M. (1988) Chilled and cryogenic preservation of tilapia spermatozoa. M.Sc. thesis, Institute of Aquaculture, University of Stirling.

Mylonas, C.C., Zohar, Y., Richardson, B.M. & Minkkinen, S.P. (1995) Induced spawning of wild American shad *Alosa sapidissima* using sustained administration of gonadotropin-releasing hormone analog (GnRHa). *Journal of the World Aquaculture Society*, **26** (Suppl. 3), 240–251.

Roberts, D.E. (1990) Photoperiod/temperature control in the commercial production of red drum *Sciaenops ocellatus* eggs. In: *Red Drum Aquaculture* (eds G.W. Chamberlain, R.J. Miget & M.G. Haby), pp. 35–43. Texas A&M University.

Roderick, E.E., Santiago, L.P., Garcia, M.A. & Mair, C.G. (1996) Induced spawning in *Oreochromis niloticus* L. In: *The Third International Symposium on Tilapia in Aquaculture* (eds R.S.V. Pullin, J. Lazard, M. Legendre, J.B. Amon Kottias & D. Pauly), p. 548. ICLARM, Makati City, Philippines.

Rottmann, R.W., Shireman, J.V. & Chapman, F.A. (1991a) *Hormone preparation, dosage calculation, and injection techniques for induced spawning of fish*. SRAC Publication no. 425. Southern Regional Aquaculture Center, Stoneville, Mississippi.

Rottmann, R.W., Shireman, J.V. & Chapman, F.A. (1991b) *Techniques for taking and fertilizing the spawn of fish*. SRAC Publication no. 426. Southern Regional Aquaculture Center, Stoneville, Mississippi.

Silverstein, J.T., Bosworth, B.G. & Wolters, W.R. (1999) Evaluation of dual injection of LHRHa and the dopamine receptor antagonist pimozide in cage spawning of channel catfish *Ictalurus punctatus*. *Journal of the World Aquaculture Society*, **30** (Suppl. 2), 263–268.

Smith, T.I.J. & Jenkins, W.E. (1984) Controlled spawning of F_1 hybrid striped bass (*Morone saxatilis* × *M. chrysops*) and rearing of F_2 progeny. *Journal of the World Mariculture Society* **15**, 147–161.

Sneed, K.E. & Clemens, H.P. (1960) *Use of fish pituitaries to induce spawning in channel catfish*. US Fish and Wildlife Service Special Scientific Report – Fisheries, no. 329. 12 p. Department of the Interior, Washington D.C.

Stevens, E. (1966) Hormone induced spawning of striped bass for reservoir stocking. *The Progressive Fish-Culturist*, **28**, 19–28.

Tan-Fermin, J.D., Pagador, R.R. & Chavez, R.C. (1997) LHRHa and pimozide-induced spawning of Asian catfish *Clarias macrocephalus* (Gunther) at different times during an annual reproductive cycle. *Aquaculture*, **148** (Suppl. 4), 323–331.

Thomas, P. & Boyd, N.W. (1989) Dietary administration of a LHRH analog induces successful spawning of spotted seatrout (*Cynoscion nebulosus*). *Aquaculture*, **80**, 363–370.

Treece, G.D. & Fox, J.M. (1993) *Design, Operation and Training Manual for an Intensive Culture Shrimp Hatchery*. TAMU-SG-93-505, Texas A&M University Sea Grant College Program, Galveston.

Vaca, A.A. & Alfaro, J. (2000) Ovarian maturation and spawning in the white shrimp, *Penaeus vannamei*, by serotonin injection. *Aquaculture*, **182** (Suppl. 3–4), 373–385.

Watanabe, W.O., Ellis, S.C., Ellis, E.P. *et al.* (1995) Progress in controlled breeding of Nassau grouper (*Epinephelus striatus*) broodstock by hormone injection. *Aquaculture*, **138** (Suppl. 1–4), 205–219.

Woods, L.C., Kohler, C.C., Sheehan, R.J. & Sullivan, C.V. (1995) Volitional tank spawning of female striped bass with male white bass produces hybrid offspring. *Transactions of the American Fisheries Society*, **124** (Suppl. 4), 628–632.

Young, M.J.A., Fast, A.W. & Olin, P.G. (1986) Induced spawning of the Chinese catfish *Clarius fuscus* in Hawaii. *Book of Abstracts: Aquaculture '86 – Reno*, World Aquaculture Society, Baton Rouge.

Chapter 10
Transgenic Aquatic Organisms

10.1 INTRODUCTION

One approach to genetic improvement of aquatic organisms that has emerged as a discipline in its own right in recent years is transgenesis, the transfer of foreign genes into new hosts. Transgenic fish (or mollusks or crustaceans) can be defined as possessing within their chromosomal DNA, either directly or through inheritance, genetic constructs which have artificial origins. The key word for researchers here is *within* the chromosomal DNA: introduced constructs must be incorporated into the target organism in such a way as to be expressed and passed along to subsequent generations.

The potential pay-offs for utilizing this type of technology in aquaculture are high: rapid, almost instantaneous gains in many types of important production traits such as growth, cold tolerance, or disease resistance may be possible. The potential problems, however, also are impressive: labor- and capital-intensive methodologies and consumer distrust of genetically engineered products in some nations. Another major constraint to widespread adoption of transgenic stocks in aquaculture involves regulatory restrictions on stocking and culture of genetically modified organisms. Due to a lack of performance data, it is difficult to assess (or even speculate on) the potential impacts of genetically modified aquatic organisms on natural systems. As a result, resource managers are reluctant even to attempt to develop protocols for the use of these organisms in situations where inadvertent releases could occur.

Transgenic stocks may soon be available to aquaculturists in several regions of the globe. Accordingly, a discussion of their origins and characteristics is probably in order. In order to address concerns relating to the potential environmental impacts of transgenic organisms that might escape from aquaculture facilities, many researchers are investigating the use of polyploidy or sex control to prevent reproduction in transgenic stocks. This also makes sense from a commercial standpoint. While our understanding of the performance of transgenic aquatic organisms improves with each study published, the techniques employed are also evolving. Components of various viruses, lipofection, microparticle bombardment and other techniques are all being investigated for their utility in gene transfer as this book goes to press.

10.2 THEORY

An introduced genetic construct typically includes a structural gene, the encoded DNA instructions for the production of a specific protein or similar gene product. Recall, however, that such DNA is nothing more than instructions, serving as a template from which proteins are assembled. For these foreign genes to be successfully expressed in the recipient organism, the introduced constructs must not only become physically incorporated on a target chromosome, they must also contain genetic sequences that serve as promoters and terminators for their transcription. In recent decades, researchers have introduced many structural "genes" into aquatic organisms. Some of these have come from fish such as trout, salmon or ocean pout, but others have originated in cows, birds, insects, bacteria and even viruses. Many different promoter sequences have also been evaluated, as have elements to facilitate insertion into receiving chromosomes.

10.3 PRACTICE

Numerous variables come into play in the production of transgenic organisms, such as the number of construct copies incorporated, where and on which chromosomes incorporation takes place, epistatic interactions with alleles at other loci, or potential linkage to sex-determining or influencing loci. Often, researchers have determined that it takes one or several cycles of cell division before any copies of a construct connect with and are spliced into chromosomal DNA strands. When only a portion of an organism's cells contain the introduced construct, the resultant animal is referred to as a mosaic. Under these circumstances successful transgenesis is still possible as long as some of the cells that will eventually form ovaries or testes receive copies of the construct.

Additionally, incorporation does not necessarily mean a new "gene" will be expressed. As a result of the effects of normally occurring regulatory sequences throughout the receiving DNA, some sites appear to be much more conducive to gene expression than others. These differences are apparent both on and among chromosomes. In any case, transgenic (or potentially transgenic) organisms must be raised to maturity and allowed to reproduce in order to determine if the introduced construct can be successfully passed along to subsequent generations.

10.3.1 Microinjection

Apart from the actual assembly of the genetic constructs (which is beyond the scope of this discussion), the process of gene transfer is surprisingly straightforward. Early transgenic research on aquatic animals relied on microinjection of genetic constructs. An extremely fine glass tube was typically utilized to introduce multiple copies of constructs directly into newly fertilized eggs. Ideally, at least one or a few introduced constructs would find their way onto a chromosome immediately after the host egg had been fertilized and before the first cell division. In those instances when this was in fact achieved, the construct(s) would generally be replicated with each cell division, and all the cells in the developing organism, including future gonads, would contain copies.

10.3.2 Electroporation

In recent years, production of transgenic aquatic organisms has also been achieved through a process called electroporation. In this approach, fertilized ova are placed in a solution containing millions of copies of the genetic construct of interest. High voltages of extremely short duration are then used to produce a transient, high-permeability state in the membranes of the ova. The temporary permeability allows for constructs to cross into the ova, and the voltage can, in theory, simultaneously facilitate transport of the construct molecules through the membrane. An alternative electrical approach has also been reported. It involves use of thin-film electrodes on a glass plate to apply a localized electric field to the animal poles of fertilized eggs.

10.3.3 Biolistics

The term biolistics was coined to reflect the combined concepts of ballistics and biology. Biolistics involves the firing of microscopic particles coated with DNA constructs directly into living cells, through the use of devices known as gene guns. The particles themselves, usually made of gold or tungsten, carry the DNA into the cell interior where it can subsequently be incorporated into the host genome. A variety of particles has been evaluated for this use, with different sizes, shapes and chemical compositions. One particularly interesting design involves donut-shaped tungsten particles, which provide some degree of protection for an internal reservoir of constructs during the delivery process.

10.3.4 Lipofection

The process of lipofection, or liposome-mediated transfection, has been increasingly important in production of transgenic aquatic organisms in recent years. Lipofection, which is widely used in transformation of cell culture lines for experimental purposes, involves the encapsulation of DNA constructs in lipid vesicles and subsequently bringing the vesicles into contact with a living target cell. As a lipid vesicle contacts a cell, uptake can take place through fusion with the plasma membrane and/or endocytosis. Lipofection reagents can occasionally be toxic to target cells and may alter cellular function or select for specific cell types.

10.3.5 Incorporation and integration

It has been estimated that only about one construct in several hundred thousand to a million is ever successfully incorporated to produce a transgenic organism, so eggs are usually exposed to millions of copies in a shotgun approach. On an individual organism basis, disastrous consequences can result from this approach depending on where a construct is incorporated within the existing genetic code. Introduced sequences of DNA have the potential to alter or completely disable major genes and the proteins they normally produce, or to change relationships among genes or traits they influence.

This phenomenon of random incorporation also leads to problems in developing true-breeding lines of transgenic fish, reflected in the need to produce individuals with the introduced construct in the same location or locations on each of their homologous chromosome sets. This is referred to as "integration" into the host genome.

10.4 ILLUSTRATIVE INVESTIGATIONS AND APPLICATIONS

10.4.1 Transgenesis in Indian catfish

Sheela *et al.* (1999) reported on the successful transfer and integration of two distinct genetic constructs containing growth hormones into the Indian catfish *Heteropneustes fossilis* using electroporation. Each of these constructs also included additional base-pair sequences. A promoter sequence commonly referred to as Zp was integrated into both constructs, as was a reporter gene. Reporter genes produce specific products which can be observed or detected through various assays to indicate if genetic constructs have been incorporated, integrated, and expressed in the target organisms.

Freshly fertilized eggs were electroporated at conditions of voltage, capacitance resistance and timing which had already been determined. Survival at hatching averaged 56% for electroporated eggs, compared to 70% for controls. Although the treatment itself may have reduced survival, the potential alteration of major genes by random construct incorporation in many ova may also have contributed to the observed increase in mortality.

Presumed transgenic fish (based on the reporter gene, β-galactosidase) exhibited growth increases ranging from 30% to 60%. However, transgenic (or potentially transgenic) organisms must be raised to maturity and allowed to reproduce in order to determine if the introduced construct can be successfully passed along to subsequent generations. In this study, transgenes were indeed inherited in many progeny from transformed fish, as determined both through DNA analyses and expression of the reporter gene.

10.4.2 Transgenic tilapia

Rahman and Maclean (1999) produced three lines of transgenic tilapia using a construct built with a chinook salmon (*Oncorhynchus tshawytscha*) growth hormone and a regulatory sequence from ocean pout (*Macrozoarces americanus*). The transmission rate of these new genes from the first transgenic generation (G_0) to their offspring (G_1) was less than 10%, reflecting the presence of the transgene in only a small portion of the (G_0) germ cells. Nonetheless, subsequent transmission from G_1 to G_2 followed expected Mendelian ratios.

The chinook growth hormone was clearly expressed in all transgenic tilapia produced: the average weight of G_1 and G_2 fish was roughly 300% that of non-transgenic controls. In a related study (Rahman *et al*. 1998), however, transgenic tilapia possessing a construct incorporating a growth hormone gene and promoter from sockeye salmon exhibited no improvement in growth even though the same construct enhances growth significantly in salmon.

Martinez *et al*. (1999) reported on separate ongoing work involving transgenic tilapia. They produced transgenic tilapia (*Oreochromis hornorum urolepis*) by microinjecting developing eggs with constructs containing tilapia growth hormone and regulatory material derived from a human cytomegalovirus. Subsequently, they identified a male, containing one copy of the transgene, to serve as founder of a transgenic line of tilapia. The transgene was inherited in a simple Mendelian fashion, indicating stable, permanent transformation of the germ-line. Second-generation transgenic fish were significantly larger than their non-transgenic siblings, and a transgene dosage effect (homozygous vs. heterozygous) appeared to be present.

10.4.3 Atlantic salmon

Fast growth is not the only benefit that can be derived from the application of gene transfer in aquaculture. Sometimes, this is true even when the genes being transferred control growth. Saunders *et al*. (1998) reported on the performance of growth-hormone transgenic Atlantic salmon (*Salmo salar*). They also used a construct built with a chinook salmon (*Oncorhynchus tshawytscha*) growth hormone and a regulatory sequence (antifreeze gene promoter) from ocean pout (*Macrozoarces americanus*). At 5 months of age, in the month of June, G_2 fish were approximately 16 cm in length and most survived >96 h following direct transfer to 35 ppt sea water. In contrast, their non-transgenic siblings were <10 cm in length and exhibited 100% mortality upon transfer to sea water.

Interestingly, treatments of high temperature (16°C) and constant light which would normally inhibit successful smoltification in normal fish had little effect on smolt development or post-smolt survival in transgenic fish. Temperature and photoperiod regimes that might promote faster early growth but inhibit normal smolt development appear to be well-tolerated by these transgenic fish. At least one down-side has been reported for growth hormone transgenic Atlantics, though: Stevens *et al.* (1998) reported that while transgenic *S. salar* grew 2–3 times faster than control fish, they exhibited a 60–70% increase in oxygen uptake under routine culture conditions and during forced swimming. In a follow-up report, however, Cook *et al.* (2000b) determined that when integrated over time from first feeding to smolt size, the transgenic salmon actually consumed 42% less total oxygen than their nontransgenic controls.

Cook *et al.* (2000a) reported growth increases of 262–285% in F_2 transgenic Atlantic salmon when compared to normal controls during early growth (up to 55 g). In addition to better growth, transgenics exhibited a 10% improvement in feed conversion efficiency, which would ultimately serve to further reduce environmental impacts of salmon production in comparison to domestic livestock.

10.4.4 Carp

Occasionally, somewhat inconsistent results may arise when dealing with transgenic fish. Hinits and Moav (1999), however, illustrate the importance of evaluating the influence of the production environment on the performance of the fish in question. Working with common carp (*Cyprinus carpio*) that had received growth hormone gene constructs (common carp β-actin proximal promoter fused to chinook salmon growth hormone DNA), they found that summer culture in earthen ponds resulted in no significant improvement in growth over control fish. However, under winter conditions transgenic fish with the same construct exhibited significantly higher growth rates than their controls, as did another group with a similar transgene.

Contrary to the impression that may have been conveyed previously, eggs are not the only potential target for electroporation. Kang *et al.* (1999) described the use of electroporation on carp sperm, and provided some insights into improving the efficiency of this approach. They examined the use of an osmotic differential to encourage the uptake of DNA-bearing medium into carp (*Cyprinus carpio*) sperm during electroporation.

As in many freshwater species, when carp sperm are exposed to a hyposmotic (less salty) solution, like fresh water, they become motile. Since these sperm are very short-lived, there is little opportunity to try to encourage their uptake of foreign DNA before they have exhausted their motility. The authors, however, dehydrated the carp sperm using a hyperosmotic (more salty) solution and subsequently rehydrated them to their original osmotic state using a solution containing genetic constructs of interest.

During rehydration in the DNA-containing solution, the researchers also applied electroporation. Gene transfer success was estimated at 66% for the electroporation–rehydration protocol, compared to only 20% for electroporation at isosmotic (stable) conditions. It would appear that this approach can significantly improve the uptake of foreign DNA into sperm cells, while preventing the onset of motility.

10.5 REFERENCES

Cook, J.T., McNiven, M.A., Richardson, G.F. & Sutterlin, A.M. (2000a) Growth rate, body composition and feed digestibility/conversion of growth-enhanced transgenic Atlantic salmon (*Salmo salar*). *Aquaculture*, **188** (Suppl. 1–2), 15–32.

Cook, J.T., McNiven, M.A. & Sutterlin, A.M. (2000b) Metabolic rate of pre-smolt growth-enhanced transgenic Atlantic salmon (*Salmo salar*). *Aquaculture*, **188** (Suppl. 1–2), 33–45.

Hinits, Y. & Moav, B. (1999) Growth performance studies in transgenic *Cyprinus carpio*. *Aquaculture*, **173** (Suppl. 1–4), 285–296.

Kang, J.H., Yoshizaki, G., Homma, O., Strussman, C.A. & Takashima, F. (1999) Effect of an osmotic differential on the efficiency of gene transfer by electroporation of fish spermatozoa. *Aquaculture*, **173** (Suppl. 1–4), 297–307.

Martinez, R., Arenal, A., Estrada, M.P. *et al.* (1999) Mendelian transmission, transgene dosage and growth phenotype in transgenic tilapia (*Oreochromis hornorum*) showing ectopic expression of homologous growth hormone. *Aquaculture*, **173** (1–4), 271–283.

Rahman, M.A. & Maclean, N. (1999) Growth performance of transgenic tilapia containing an exogenous piscine growth hormone gene. *Aquaculture*, **173** (Suppl. 1–4), 333–346.

Rahman, M.A., Mak, R., Ayad, H., Smith, A. & Maclean, N. (1998) Expression of a novel piscine growth hormone gene results in growth enhancement in transgenic tilapia (*Oreochromis niloticus*). *Transgenic Research*, **7** (Suppl. 5), 357–369.

Saunders, R.L., Fletcher, G.L. & Hew, C.L. (1998) Smolt development in growth hormone transgenic Atlantic salmon. *Aquaculture*, **168** (Suppl. 1–4), 177–194.

Sheela, S.G., Pandian, T.J. & Mathavan, S. (1999) Electroporatic transfer, stable integration, expression and transmission of pZpssypGH and pZpssrtGH in Indian catfish, *Heteropneustes fossilis* (Bloch). *Aquaculture Research*, **30** (Suppl. 4), 233–248.

Stevens, E.D., Sutterlin, A. & Cook, T. (1998) Respiratory metabolism and swimming performance in growth hormone transgenic Atlantic salmon. *Canadian Journal of Fisheries and Aquatic Science*, **55** (Suppl. 9), 2028–2035.

Chapter 11
Genetic Threats to Wild Stocks and Ecosystems

11.1 INTRODUCTION

An occasional criticism of aquaculture in many parts of the world is the potential for escaped production stocks to impose negative genetic and ecological impacts on wild populations, both within and among species. Clearly, fish culturists must become more aware of genetic issues involved with escaped stocks. Around the globe, concern is growing over genetic conservation of wild populations of aquatic species – not only in terms of genetic variation within isolated populations, but also *among* populations of any given species. The later issue is more concerned with future evolutionary potential than current efforts to maintain species survival, but it has begun to play an important role in shaping conservation genetics policy for many aquatic species.

11.2 THEORY

A brief review is probably in order concerning the types of genetic risks facing wild populations when they are exposed to escapees from aquaculture operations. Perhaps the most commonly cited threat is loss of population identity. Indeed, a number of direct effects from aquacultured stocks can ultimately cause changes in the genetic structure of a wild population. Perhaps the most dramatic is sometimes referred to as "swamping," when isolated populations are faced with overwhelming numbers of introduced individuals. There is a general tendency to look at the genetic make-up of populations in terms of gene frequencies, and this makes good sense within a population of relatively constant size. But raw numbers come into play when examining gene frequencies for a mixed population of introduced and native fish.

Though less dramatic, perhaps a more real threat than swamping is the continuous introduction of small numbers of farmed breeding stock into wild populations. The introduction of genetic material in this way is sometimes termed "introgression." Hayes *et al.* (1996) examined native, stocked and hybrid populations of brook trout in the southern Appalachian mountains in North America and determined that introgression from continued introductions of non-native strains could ultimately eliminate unique genotypes within the species. Introgression has also become a concern for many stocks in the North

American *Morone* complex, where government-stocked hybrids can occasionally back-cross with parental species and introduce previously foreign genetic material into established populations.

Escapes and introductions of cultured fish in low numbers might be considered as a phenomenon similar to "straying" in salmonid populations – where some individuals find their way to spawning grounds other than those they originated from. This is clearly recognized as a way for many relatively isolated populations to maintain genetic diversity in the face of limited population size, genetic drift, and inbreeding. On the other hand, too much of this type of exchange can undermine genetic variation *among* populations, the important species attribute that is earning respect under conservation genetics initiatives around the globe. An argument has often been made, however, that when resident populations have already declined due to exploitation and/or habitat alteration, introduced fish may hold out the promise of increased genetic diversity to allow for re-adaptation of the remnant population. While this can sometimes be the case, the new variation produced will almost certainly include alleles that were not previously present, and in such a situation it must be accepted that the former wild population has already ceased to exist.

Even if natural populations are reproductively isolated from aquaculture escapees, the need to accommodate the presence of a new species may inevitably force changes in native populations through selection. Such changes are due to competitive pressures from introduced fish and/or their offspring. An often-correlated impact is reduced within-population variation. The presence of competing aquacultured escapees can also result in an effective reduction in the size of the wild population, even if it maintains its genetic integrity by reproductive isolation. These reductions would in turn increase chances of inbreeding and of drift.

11.3 ILLUSTRATIVE INVESTIGATIONS

The tools available to examine genetic variation both within and among populations of fish are improving all the time, but the state of the art may still leave something to be desired. An excellent example of this was presented by Chapman (1989). He compared two types of genetic material, mitochondrial DNA (mtDNA) and forms of AAT genes found in the nuclear (chromosomal) DNA, as indicators of genetic divergence among introduced populations of bluegills (*Lepomis macrochirus*). Analyses of three pond populations of bluegill suggested significant mtDNA divergence 17 years after they were established from a common hatchery population. However, no differentiation was apparent in terms of nuclear DNA. Hansen and Loeschcke (1996) pointed out similar dissimilarities between mtDNA and nuclear gene frequency dynamics in brown trout.

Consider differential selection pressures with regard to mtDNA as well as different effective population sizes. Normally, mtDNA is inherited uniparentally, only from the mother through the egg contents, as compared to nuclear DNA. In a study with rainbow trout (*Oncorhynchus mykiss*), Danzmann and Ferguson (1995) reported that fish with two specific mtDNA forms were significantly larger and had higher condition factors than fish with two distinctly different mtDNA forms. These phenomena could interact to drive mtDNA frequencies in entirely different directions from nuclear DNA frequencies over time. The use of mtDNA as an indicator of population divergence or differentiation requires an entirely different set of assumptions than might be appropriate for nuclear DNA.

In another example of discrepancies in genetic characterization of populations, Skaala (1994) used biochemical genetic markers (nuclear DNA gene forms for various enzyme systems) and a visible genetic marker to examine gene flow from introduced farmed individuals to native brown trout (*Salmo trutta*) populations. The visible marker was a variant of the normal brown trout spotting pattern, where homozygotes (individuals with both copies of the variant gene) and heterozygotes (individuals with only one copy) could be identified. Genetically marked spawners were released into two sections of a small river, above and below a waterfall that prevented upstream movement of stocked or native trout. In each section, the number of released fish exceeded the number of native fish present.

One year later, the genetic contribution of the introduced fish in the upstream section was calculated as 19.2% based on the biochemical marker and only 12.8% based on the visible marker. Three years after stocking, their estimated contribution had declined to 10.8%. No homozygotes for the spotting pattern were found. In the downstream section below the falls, the visible marker was only observed in the heterozygous (single-copy) state. No homozygotes were ever recovered. One year after stocking the estimated genetic contribution of the introduced fish was 7.5% based on the visible marker and 16.3% based on the biochemical marker. These values dropped to 2.7% and 10.9%, respectively by 17 months post-stocking. In light of the fact that some reproduction occurred among introduced fish, this would tend to negate the often-heard but generally unfounded claim that cultured fish escaping at an early life stage may be more adaptable to wild habitats and therefore pose a greater threat to wild populations.

These results should come as no surprise, upon reviewing other sound science on this topic. In the case of walleye (*Stizostedion vitreum*) populations in Saginaw Bay, Michigan, a similar pattern has been reported. Todd and Haas (1995) indicated that in spite of intensive stocking of walleye beginning in 1979, subsequent population recoveries in the lake appeared to be based on reproduction by the endemic population. While allele frequencies in the 1983 spawning run were similar to those of the pond-reared walleyes that had been stocked into the lake, subsequent trends indicate that reproduction by genotypes other than stocked fish has greatly exceeded any genetic contribution from the hatchery stock.

Hansen and Loeschcke (1996) reported that the population of brown trout in a Danish river consisted almost entirely of indigenous trout, in spite of heavy stocking with hatchery-reared brown trout up until 1990. Previously, Skaala (1994) pointed out that while the number of wild Atlantic salmon spawners in Norwegian rivers had been estimated at roughly 100 000 fish, the number of salmon escaping from farms in that country's waters had reached about 2 million in 1989. And yet, wild populations were still identifiable and distinct. Perhaps much of what we consider genetic improvement in aquaculture stocks is not really improvement at all under natural conditions.

11.4 REFERENCES

Chapman, R.W. (1989) Mitochondrial and nuclear gene dynamics of introduced populations of *Lepomis macrochirus*. *Genetics*, **123**, 399–404.

Danzmann, R.G. & Ferguson, M.M. (1995) Heterogeneity in the body size of Ontario culture rainbow trout with different mitochondrial DNA haplotypes. *Aquaculture*, **121** (Suppl. 4), 301–312.

Hansen, M.M. & Loeschcke, V. (1996) Temporal variation in mitochondrial DNA haplotype frequencies in a brown trout (*Salmo trutta* L.) population that shows stability in nuclear allele frequencies. *Evolution*, **50** (Suppl. 1), 454–457.

Hayes, J.P., Guffey, S.Z., Kriegler, F.J., McCracken, G.F. & Parker, C.R. (1996) The genetic diversity of native, stocked and hybrid populations of brook trout in the southern Appalachians. *Conservation Biology*, **10** (Suppl. 5), 1403–1412.

Skaala, O. (1994) Measuring gene flow from farmed to wild fish populations. In: *OECD Environment Monographs, 90. Ottawa '92: The OECD workshop on methods for monitoring organisms in the environment*. Environment Directorate, Organisation for Economic Co-operation and Development, Paris.

Todd, T.N. & Haas, R.C. (1995) Genetic contribution of hatchery fish to walleye stocks in Saginaw Bay, Michigan. *American Fisheries Society Symposium*, **15**, 573–574.

Index

6-DMAP, 130, 148, 154–5
17,20-P, 186, 209
Abramis brama, 184
Acanthopagrus latus, 185, 190
Acipenser baeri, 112
Acipenser guldenstadti, 112
Acipenser ruthenus, 112
additive genetic effects, 34, 36–7, 42, 62, 64, 67–9, 80, 99
Aeromonas salmonicida, 63, 109
  see also furunculosis
albinism, 22, 25–30, 44, 126, 129–30
allele, 4–6, 8, 10–11, 35–6, 94–5, 219, 227
Alosa sapidissima, 206
amino acids, 5
analysis of variance (ANOVA), 37, 39–40, 44, 52, 54, 61, 81
anchovies, northern, *see Engraulis mordax*
androgenesis, 25, 120, 124–132, 13–137
androgenic gland, 164, 170
ANOVA, *see* analysis of variance
antibiotics, in sperm diluent, 196
antibody production, 63–4, 87, 136
antifreeze gene, 222
Arctic charr, *see Salvelinus alpinus*
Argopecten irradians concentricus, 88
Argyrosomus hololepidotus, 210
Aristichthys nobilis, 190
artificial fertilization, 197–8, 211
assortative mating, 55, 57, 83
astaxanthin, 29
asynchronous ovulation, 59, 206–7
Atlantic croader, *see Micropogonias undulatus*
Atlantic salmon, *see Salmo salar*
autosomal loci, 19–21, 134, 163, 166
average effects of alleles, 35–6, 40, 43
ayu, *see Plecoglossus altivelis*

Bairdiella icistia, 198–9
Barbus longiceps, 104
base pairs, 4–5
bass, Australian, *see Macquaria novemaculeata*
bass, largemouth, *see Micropterus salmoides*
bass, sea, European, *see Dicentrarchus labrax*
bass, sea, white, *see Atractoscion nobilis*
bass, smallmouth, *see Micropterus dolomieui*
bentonite, 197
Betta splendens, 9, 127, 143
bighead carp, *see Hypophthalmichthys nobilis*
biolistics, 220
bivalves, 220
black drum, *see Pogonias chromis*
bluegill, *see Lepomis macrochirus*
BLUEs, 51
BLUPs, 51
Brachydanio rerio, 131
bream, sea, *see Abramis brama*
bream, sea, black, *see Acanthropagrus schlegeli*
bream, sea, gilthead, *see Sparus aurata*
bream, sea, red, *see Pagrus major*
breed formation, 98, 105–6
breeding programs, 2, 34, 51, 66, 89, 93
breeding value, 34, 62, 69, 70–71, 79–80
broodstock
  collection, 106, 179
  conditioning, 59, 174, 179, 192–3
  maintenance, 100, 166, 174, 179, 194
brook charr, *see Salvelinus fontinalis*
brook trout, *see Salvelinus fontinalis*
brown trout, *see Salmo trutta*
Bufo regularis, 184

Index

bullhead, black, *see Ictalurus melas*
bullhead, brown, *see Ictalurus nebulosus*
bullhead, yellow, *see Ictalurus natalis*

Cancer irroratus, 30
canthaxanthin, 54
Capoeta damascina, 104
carotenoids, 54
carp, 105
carp, bighead, *see Hypophthalmichthys nobilis*
carp, black, *see Mylopharyngodon piceus*
carp, common, *see Cyprinus carpio*
carp, grass, *see Ctenopharyngodon idella*
carp, koi *see* koi; *Cyprinus carpio*
carp, leather, 16
carp, mirror, 16–17, 109, 136
carp, silver, *see Hypophthalmichthys molotrix*
Carassius auratus (goldfish), 27, 121, 136
catfish, African, *see Clarias anguillaris; Clarias gariepinus*
catfish, Asian, *see Clarias batrachus*
catfish, blue, *see Ictalurus furcatus*
catfish, channel, *see Ictalurus punctatus*
catfish, Chinese, *see Clarias fuscus*
catfish, European, *see Silurus glanis*
catfish, flathead, *see Pylodictus olivarius*
catfish, white, *see Ictalurus catus*
Catla catla, 25–6, 102, 104, 190
Catostomus commersoni, 210
centromere, 125, 133
Chanos chanos, 199
Cherax destructor, 211
Cherax quadricarinatus, 62, 90
Cherax tenuimanus, 89
Cherax spp., 171
chicken, 1, 184
chromosomes
  homologous, 4, 7, 221
  pairing, 4, 111–2, 123, 126–7, 141–2
  replication, 5–8, 120–21, 123, 127, 136
  sex, 4, 70, 134, 163
  structure, 2, 4–5, 105, 125–6, 133, 219
  translation, 5
Cirrhinus mrigala, 190
clam, dwarf surf, *see Mulinia lateralis*
clam, giant, *see* tridacnids
clam, Manila, *see Ruditapes philippinarum*

Clarias anguillaris, 114
Clarias batrachus, 114, 207
Clarias fuscus, 143, 190, 199
Clarias gariepinus, 27, 114, 129
Clarias isheriensis, 184
Clarias macrocephalus, 111–2, 134, 208
clones, 120, 123–5, 135, 137, 167
coloration, 2, 4, 6, 9
  inheritance, 10–25, 29, 33, 35, 54, 61–2, 101, 126, 162
  characterizing, 6, 18–20, 22–4
combining ability, 63, 99
common carp, *see Cyprinus carpio*
continuous traits or characters, 33
conservation, genetic, 225–6
correlations
  genetic, 61–2, 73–4, 83–4
  phenotypic, 24, 61, 63–4, 85
cortisol, 63, 193
Corydoras spp., 27
covariance, 37–41, 55, 57, 62, 68–9, 73
CPH, 207
Crassostrea gigas, 115, 135, 143, 154–7
Crassostrea virginica, 115, 155
crayfish, 28–9, 34, 44–6, 57–9, 62–3, 85, 89–90, 102, 111, 115, 171, 192, 211
croaker, *see Bairdiella icistia*
croaker, Atlantic, *see Micropogonias undulatus*
crossbreeds and crossbreeding, 34, 43, 82, 93, 98–102, 104–110, 112, 115
crossing over, 7, 127, 134
crustaceans, 44, 49, 58, 84, 89, 164, 170, 192, 211
Ctenopharyngodon idella, 26, 105, 111, 130, 143, 147, 153, 190
culling, 74, 85
Cynoscion nebulosus, 193
cyprinids, 25, 149–50, 152, 192
Cyprinus carpio, 17, 19, 87, 108, 126, 128–9, 131, 133–7, 143, 150, 165, 197, 210, 223
cytochalasin B, 127, 130, 135–6, 148, 154–7
cytoplasmic inheritance, 38, 63, 102

dams, 39–41, 44
daylength, *see* photoperiod
diallel cross, 52, 63–4, 113
Dicentrarchus labrax, 143, 154, 185, 190

diet, 45, 79, 169, 193
diethylstilbestrol, 167
diploid, 4, 7, 105, 120, 123–4, 141, 144, 157
directional dominance, 98
disease resistance, 16–7, 52–3, 63–4, 101
DNA, 5–7, 9, 121, 218–21, 226–7
domestication, 81
dominance, 6, 12, 35
dominance deviations, 35–6, 94
dominance genetic effects, 34–6, 41, 43, 56, 59, 62, 67–8, 93–5, 98–9, 109
domperidone, 185, 207, 210
dopamine, 177
   antagonists or blockers, 185, 188, 207–8, 210
dressout percentage, 51, 56, 60, 63, 69, 76–7, 84–5, 104
drift, 75, 96–7, 226
drum, black, see Pogonias chromis
drum, red, see Sciaenops ocellatus

Echinometra spp., 115
effective population number or size, 78, 95–6, 98, 100, 226
eggs, 8
   adhesive, 197
electrofusion, 154
electroporation, 220, 221, 223
Engraulis mordax, 198
environmental conditioning and maturation, 45, 174–6, 180
environmental tolerances, 37, 85–6, 101
Epinephelus aeneus, 211
Epinephelus striatus, 207
epistasis, 24, 34, 38, 75, 167, 219
Esox lucius, 115
Esox masquinongy, 115
ethynylestradiol, 167–8
extenders, 197
eyestalk ablation/enucleation, 211–2

fathead minnow, 18
fecundity, 2, 33, 53, 69, 75–6, 84, 100–101
feminization, 165, 167–9
fertilization, 9, 115, 120–21, 148, 194, 197, 220
fixation of alleles, 95, 97–8, 106, 169
fixed effects, in variance analysis, 50–1, 86

flounder, southern, see Paralichthys lethostigma
food conversion, 25, 60, 223
full siblings or full-sibs, 37–8, 40–41, 76–7
Fuller's earth, 197
furunculosis, 63
   see also Aeromonas salmonicida

gametes, 6–7, 35, 41, 44, 94–5, 100, 104, 121, 124–5, 141, 163–4, 176, 193
gene, 2
gene transfer, 218–223
genetic constructs, 218–223
genetic drift, 75, 96–7, 226
genetically modified organisms, 218
genotype-environment interactions, 86
genotypic ratios, 12, 222
GIFT program tilapia, 81, 83
GnRH (gonadotrophin releasing hormone), 176–7, 185
GnRHa (GnRH analogue), 59, 185–6
goldfish, 27, 121, 136
gonadosomatic index, 134–5
gonadotropic hormones, 177, 183
gonadotropins, 176–7, 181, 183–4, 189
grading, 88
grass carp, see Ctenopharyngodon idella
grouper, Nassau, see Epinephelus striatus
grouper, white, see Epinephelus aeneus
groupers, 207, 211
growth hormone, 221–3
GtH's, 177, 183–5
guppy, see Poecilia reticulata
gynogenesis, 120–31

half siblings or half-sibs, 37, 40–42
haloperidol, 185, 188
Haplochromis spp., 22
haploid, 7–8, 120–21, 123
HCG, 180, 183–6, 188
heritability, 43, 51–3
   broad sense, 68
   narrow sense, 68
   realized, 66, 71
heterogametic genders, 164–5
heterogenous growth, 170
Heteropneustes fossilis, 190, 221
Heteropneustes longifilis, 114
heterosis, 34, 93–5, 98, 101–2

heterozygous, 6, 11–2, 99, 106, 125, 127
hierarchies, 47, 49
homeostasis, 135
homogameity, 124, 133, 136, 164–5, 167, 169
homogametic gender, 163–8
homozygous, 6, 10–12, 15, 35, 87, 97, 124–7, 135, 166–7, 228
hormones, 124, 162, 164–9, 174–80, 183–9, 192–4, 199, 202–3, 205–12, 221
hybrid, 13, 22–5, 93, 98, 101–4, 108, 112–4, 150, 166, 171, 202–3
hybrid striped bass, *see Morone saxatilis* X *Morone chrysops*
hybrid vigor, 34, 93, 100, 114, 164
hybridization, 24, 93–4, 98, 100–104, 164, 166
Hypophthalmichthys molotrix, 105
Hypophthalmichthys nobilis, 143, 150, 191
hypothalamus, 176–7, 183, 185–6

Ictalurus melas, 27
Ictalurus punctatus, 26, 54–5, 64, 73, 77, 81, 84, 101, 133, 143, 152–3, 157, 165, 169, 208
IHNV (infectious hematopoietic necrosis virus), 53
immune response, 52, 63, 184
implants, 59, 169, 189
inbreeding, 10, 27, 71, 76, 79–80, 93–8, 226
    coefficient, 94–7
    depression, 27, 34, 75, 78, 93–4, 98–9
    effects, 87, 100–101, 169
independent culling, 74
injection procedures, 184–9
interploid triploidy, 142, 146–8, 156
introgression, 225
irradiation, 123, 125, 127, 131

Japanese charr, *see Salvelinus leucomaenis pluvius*

karyomorphology, 105, 111
karyotype, 105, 111, 141
koi, 19–21, 126, 129–30, 136

Labeo dussumieri, 209
Labeo fimbriatus, 104
Labeo rohita, 102, 191

Labeolabrax japonicus, 191
latency, 208
Lepomis macrochirus, 226
Lepomis spp., 102, 111, 115, 170
LHRH (leutinizing hormone releasing hormone), 185–6
LHRHa (leutinizing hormone releasing hormone analogue), 185–6, 188, 193
light, 102, 179, 202, 209, 212, 222
linkage, 120, 133–4, 219
lipofection, 218, 220
Litopenaeus vannamei, 89
    see also Penaeus vannamei
locus, 5–6, 8, 12, 36, 38, 94, 99, 164
Lutjanus analis, 191
Lutjanus campechanus, 191

Macquaria novemaculeata, 180
Macrobrachium acanthurus, 45
Macrobrachium nipponense, 85
Macrobrachium rosenbergii, 55–7, 84, 86, 170
Macrobrachium spp., 51, 57
Macrozoarces americanus, 222
Manila clam, *see Ruditapes philippinarum*
markers
    artificial, 40, 49, 79
    genetic, 41, 136, 153, 227
    phenotypic, 44, 49, 227
marron, *see Cherax tenuimanus*
masculinization, 124, 164–5, 169–71
maternal influences, 9, 34, 38, 41, 53, 102–3, 110, 131, 154
mating designs, 38–45, 73, 79
meiosis, 7–9, 35, 94, 120–23, 126, 128, 130, 133
melanin, 25–6, 28
Mendelian inheritance, 6, 222
Menidia menidia, 166
Mercenaria mercenaria, 83
methyltestosterone, 169
metoclopramide, 185, 209
metric traits or characters, 33–4, 36, 50, 66–7, 93, 98
microinjection, 219
Micropterus dolomieui, 199
Micropterus salmoides, 115, 191
micropyle, 195
milkfish, *see Chanos chanos*

Minytrema melanops, 210
Misgurnus anguillicaudatus, 127, 130, 148
mitochondria, 9, 41, 102, 111, 124, 156, 226
mitosis, 7, 123–5, 128–30, 145
mixed models, in variance analysis, 51
molecular genetics, 79, 95, 99, 162
mollusks, 88, 121, 148, 212
monosex stocks, 100, 102, 104, 111, 120, 125, 162, 164–7, 170
Morone chrysops, 102–3, 191, 197–8
Morone mississippiensis, 111
Morone saxatilis, 102–3, 111, 191, 194, 202
    X *M. chrysops*, 102–3, 115, 202
Morone spp., 102, 106, 111, 170, 226
mosaic, 219
Mugil cephalus, 81
Mugil platanus, 209
Mugil so-iuy, 209
Mulinia lateralis, 136, 155
mulloway see *Argyrosomus hololepidotus*
mussels, see *Mytilus*
mutation, 24, 29
Mylopharyngodon piceus, 111
Mytilus edulis, 130, 154
Mytilus galloprovincialis, 130, 154

Nishikigoi, see koi
normal distribution, 33
Notemigonus crysoleucas, 86

ocean pout, see *Macrozoarces americanus*
Odonesthes bonariensis, 143
Oncorhynchus gorbuscha, 131
Oncorhynchus kisutch, 61–2, 81, 110, 151–2
Oncorhynchus masou, 128, 151
Oncorhynchus mykiss, 28, 41, 53, 63, 83, 86, 109–11, 128, 131, 135, 143, 150–51, 154, 227
Oncorhynchus nerka, 110
Oncorhynchus tshawytscha, 28, 49, 152, 222
oocytes, 7–9, 59–60, 121, 124, 126, 142, 192
Orconectes virilis, 211
Oreochromis aureus, 23–5, 50, 105, 132, 136, 149, 165–8
Oreochromis mossambicus, 20, 22, 24, 35, 112, 166
Oreochromis niloticus, 13–4, 21–5, 35, 50, 81–2, 105, 113–4, 131–2, 136, 143, 149, 162, 165–8, 210

Oreochromis urolepis hornorum, 24, 166, 222
Oreochromis spp., 111, 167
Ostrea chilensis, 88
Ostrea edulis, 83, 157
ovary, 186, 211, 219
overdominance, 98–9, 135
ovulation, 44–5, 176–7, 184–5, 193–5, 206
oyster
    Chilean, see *Ostrea chilensis*
    eastern, see *Crassostrea virginica*
    European, see *Ostrea edulis*
    Pacific, see *Crassostrea gigas*
    rock, see *Saccostrea*

paddlefish, 193
Pagrus auratus, 191
Pagrus major, 143, 153
Pangasius sutchi, 111–2
Paralichthys lethostigma, 200, 210
pearl oyster, see *Pinctada fucata martensii*
pedigrees, 38, 94–5
Penaeus japonicus, 89, 100, 191
Penaeus stylirostris, 85, 89
Penaeus vannamei, 85, 100, 212
    see also *Litopenaeus vannamei*
Perca flavescens, 143, 146, 165, 191
Perca fluviatilis, 208–9
perch, European, see *Perca fluviatilis*
perch, yellow, see *Perca flavescens*
phenotype, 2
photoperiod, 45, 176–9, 182, 222
photothermal conditioning, 175, 178–82, 185, 192, 199–203
Physa heterostropha, 27
pike, see *Esox lucius*
pimozide, 185, 208–9
Pinctada fucata martensii, 101
PIT identification tags, 49, 79
pituitary, 177, 183–6
    extract, 183–4, 187
Placopecten magellanicus, 154, 212
Placuna placenta, 212
Plecoglossus altivelis, 134–5
Pleuronectes ferrugineus, 211
Poecilia reticulata, 17
polar bodies, 8–9, 120–27, 130, 142, 144, 148
polyploidy, 141–2, 148, 157

Pomoxis annularis, 143
pompano, *see Trachinotus carolinus*
population genetics, 95
porgy, yellowfin, *see Acanthopagrus latus*
Procambarus alleni, 29
Procambarus clarkii, 27–9, 51–2, 55, 57–9, 63, 171
Procambarus zonangulus, 29
progeny testing, 15–6, 59–60, 76, 165–9
promoters, 219
prostaglandins, 177, 185–6, 188
Puntius gonionotus, 133

qualitative traits, 4, 6, 9, 17–8
quantitative traits, 33, 94

rainbow trout, *see Oncorhynchus mykiss*
Rana adspera, 184
Rana brevipoda, 27
Rana nigromaculata, 27
random effects, 50
recessive alleles, 10–11
recombination, 133–4
red claw crayfish, *see Cherax quadricarinatus*
red drum, *see Sciaenops ocellatus*
red sea bream, *see Pagrus major*
red swamp crawfish, *see Procambarus clarkii*
regressions, 37, 67, 69–70, 78–9
releasing hormones, 176–7, 184–5, 188–9
reporter genes, 221
Rhamdella minuta, 27
RNA, 5
Ruditapes philippinarum, 157

sGnRH-a (salmon hormone analogue) or SGnRH-A, 185, 207, 210
Saccostrea commercialis, 155
Saccostrea cucullata, 89
Saccostrea glomerata, 89
Salmo salar, 61–3, 151–2, 222–3, 228
Salmo trutta, 110–11, 143, 151, 227
salmon
  Atlantic, *see Salmo salar*
  chinook, *see Oncorhynchus tschawytscha*
  coho, *see Oncorhynchus kisutch*
  masou or masu, *see Oncorhynchus masou*
  pink, *see Oncorhynchus gorbuscha*
  sockeye, *see Oncorhynchus nerka*
Salvelinus alpinus, 61, 110–11, 151
Salvelinus fontinalis, 16, 110, 133, 151
Salvelinus leucomaenis leucomaenis, 133, 151
Salvelinus leucomaenis pluvius, 151
scallops, giant sea, *see Placopecten magellanicus*
sciaenids, 193, 210
Sciaenops ocellatus, 68, 79, 175, 199, 200–202
Scophthalmus maximus, 154
sea bass, European *see Dicentrarchus labrax*
selection, 21, 66–8, 70–71, 73–90
  domestication, 62, 76, 79, 81
  combined, 78, 83
  family, 77
  index, 73–4
  indirect, 1, 63, 84
  individual or mass, 60, 66, 77, 82, 85
  multi-trait, 73
  tandem, 74
  within-family, 78
selection differential, 70, 74
selection intensity, 71, 74
selection response, 1, 24, 67, 75, 80–81
serotonin, 185, 211–3
sex
  chromosomes, 4, 163
  determination, 18, 102, 104, 120, 133, 135, 137, 155, 162–9
  heterogametic, 164–5
  homogametic, 163–8
  ratios, 25, 111, 166–7, 171
  reversal, 162, 164, 168–71
shad, American, *see Alosa sapidissima*
shrimp, penaeid, 85, 89, 211–2
Siganus guttatus, 191
silt-clay, 197
Silurus glanis, 27, 143, 208
silver barb, *see Puntius gonionotus*
sire variance, 40
smoltification, 62, 151, 222
sodium sulfite, 197
sole, Dover, *see Solea solea*
Solea solea, 133
Sparus aurata, 59–60, 79, 87, 101, 206

sperm, 76, 115, 121, 127, 156, 196
  electroporation, 223
  formation, 78, 121, 177
  irradiation, 122–3, 127, 130
  storage, 196
spotted seatrout, *see Cynoscion nebulosus*
steelhead, 28, 110
sterility, 93, 100, 104, 141–2, 149
Stizostedion canadense, 101, 115, 191
Stizostedion vitreum, 101, 115, 169, 186, 191, 209, 227
straying, 226
stress response, 63, 81, 105, 175, 179, 193, 212
striped bass, *see Morone saxatilis*
sturgeon, 78, 105–6, 108, 112, 193
sucker, spotted, *see Minytrema melanops*
sucker, white, *see Catostomus commersoni*
swamping, 225
Sydney rock oyster, *see Saccostrea commercialis*
synthetic populations, 83, 106, 108–9, 112

Tachysurus tenuispinis, 27
tannic acid, 197
tau_0, 129–30, 136
temperature, 33, 45, 69, 78, 86, 127–9, 147–8, 152, 166, 176–9, 182, 194, 198, 201–2, 211, 222
tench, *see Tinca tinca*
terminators, 219
testes, 177, 193, 219
tetraploidy, 125, 135, 142, 144–9, 154, 156–7
threshold traits, 33, 162
tilapia
  blue, *see Oreochromis aureus*
  Mossambique, *see Oreochromis mossambicus*
  Nile, *see Oreochromis niloticus*
  red, 22–5, 50, 112–3
Tinca tinca, 129, 143, 153
Trachinotus carolinus, 198
transcription, 5, 219

transgenesis, 141, 218–23
trenbalone acetate, 169
tridacnids, 192, 212
triploids, 111, 115, 136, 141–157, 167
trout
  brook *see Salvelinus fontinalis*
  brown *see Salmo trutta*
  rainbow *see Oncorhynchus mykiss*
  steelhead *see* steelhead
trout, spotted sea, *see Cynoscion nebulosus*
turbot, *see Scophthalmus maximus*

ultraviolet irradiation, 122, 128, 130–1, 136, 154–5, 212
urea, 197

variance
  environmental, 33, 37, 38, 40–2, 45, 51, 53, 75, 134–5
  genetic
    additive, 34–8, 42–3, 66, 68–9, 135
    dominance, 37–8, 40–42, 93–4
    epistatic, or interaction, 38
  maternal, 40, 42
  phenotypic, 37–9, 43, 58, 66, 68, 70, 72, 77, 107, 134
variance components, 40, 42
vundu, *see Heteropneustes fossilis*

walleye, *see Stizostedion vitreum*
white bass, *see Morone chrysops*
white-spotted charr, *see Salvelinus leucomaenis leucomaenis*
wild type, 9, 17–8, 20–23, 26
window-pane shell, *see Placuna placenta*

Xiphophorus maculatus, 113
Xiphophorus variatus, 113

yellow perch, *see Perca flavescens*
yellowfin porgy, *see Acanthopagrus latus*
YY males, 137, 165, 168–9

zebra fish, *see Brachydanio rerio*